AS/A-LEVEL YEAR 1

STUDENT GUIDE

OCR

Chemistry A

Modules 1 and 2

Development of practical skills

Foundations in chemistry

Mike Smith

PHILIP ALLAN FOR

HODDER
EDUCATION
AN HACHETTE UK COMPANY

Philip Allan, an imprint of Hodder Education, an Hachette UK company, Blenheim Court, George Street, Banbury Oxfordshire OX16 5BH

Orders

Bookpoint Ltd, 130 Milton Park, Abingdon, Oxfordshire OX14 4SB

tel: 01235 827827

fax: 01235 400401

e-mail: education@bookpoint.co.uk

Lines are open 9.00 a.m.–5.00 p.m., Monday to Saturday, with a 24-hour message answering service. You can also order through the Hodder Education website: www.hoddereducation.co.uk

© Mike Smith 2015

ISBN 978-1-4718-4396-9

First printed 2015

Impression number 5 4 3 2 1

Year 2018 2017 2016 2015

This guide has been written specifically to support students preparing for the OCR AS and A level Chemistry examinations. The content has been neither approved nor endorsed by OCR and remains the sole responsibility of the author.

Cover photo: Galyna Andrushko/Fotolia

Typeset by Integra Software Services Pvt. Ltd., Pondicherry, India

Printed in Italy

Hachette UK's policy is to use papers that are natural, renewable and recyclable products and made from wood grown in sustainable forests. The logging and manufacturing processes are expected to conform to the environmental regulations of the country of origin.

Contents

Getting the most from this book 4

About this book 5

Content Guidance

Module 1 Practical skills assessed in the written examination . . 8

  Planning 8

  Identifying practical techniques and equipment 9

  Implementing 10

  Analysis 12

  Evaluation 17

Module 2 Foundations in chemistry 19

  Atomic structure and isotopes 19

  Compounds, formulae and equations 23

  Amount of substance 24

  Acids 34

  Redox 39

  Electron structure 43

  Bonding and structure 47

Questions & Answers

Q1 to Q10 Multiple-choice questions   Questions in this section are relevant to AS Component 1 and to A-level Components 1 and 2 56

Short response questions   Questions in this section are relevant to AS Components 1 and 2 and to A-level Components 1, 2 and 3 59

  Q11 Isotopes; electron configuration; ionisation energy 59

  Q12 Moles; choice of apparatus 62

  Q13 Redox; bonding and structure 66

  Q14 Electronegativity; hydrogen bonding 68

  Q15 Planning an experiment; interpretation of results 71

  Q16 Bonding and related properties 74

  Q17 Acids, bases and ionic equations 78

  Q18 Practical skills in chemistry 82

Extended response questions   Questions in this section are relevant to AS Component 2 and to A-level Component 3 85

  Q19 Bonding 85

  Q20 Shapes of molecules 88

  Q21 Unstructured mole calculation 89

  Knowledge check answers 91

  Index 93

■ Getting the most from this book

Exam tips

Advice on key points in the text to help you learn and recall content, avoid pitfalls, and polish your exam technique in order to boost your grade.

Knowledge check

Rapid-fire questions throughout the Content Guidance section to check your understanding.

Knowledge check answers

1 Turn to the back of the book for the Knowledge check answers.

Summaries

■ Each core topic is rounded off by a bullet-list summary for quick-check reference of what you need to know.

Exam-style questions

Commentary on the questions

Tips on what you need to do to gain full marks, indicated by the icon **e**

Sample student answers

Practise the questions, then look at the student answers that follow.

Commentary on sample student answers

Find out how many marks each answer would be awarded in the exam and then read the comments (preceded by the icon **e**) showing exactly how and where marks are gained or lost.

Short response questions

(b) (i) What amount, in moles, of $MgCO_3$ was used in the experiment? [2 marks]

(ii) Calculate the volume of $2.00\,mol\,dm^{-3}$ hydrochloric acid needed to react completely with this amount of magnesium carbonate. [2 marks]

(iii) Calculate the volume of CO_2 gas that would be produced at RTP. [2 marks]

(c) When 0.50 g of the carbonates of magnesium, calcium and barium are heated, they decompose and produce an oxide and carbon dioxide.

(i) Write an equation, including state symbols, for the decomposition of one of these carbonates. [2 marks]

(ii) Sketch and label an apparatus that could be used to measure the volume of carbon dioxide evolved. Identify a source of error in your experiment and suggest how the error could be rectified. [4 marks]

(iii) Explain why each carbonate produces a different amount of carbon dioxide. [2 marks]

Total: 16 marks

e The command word 'calculate' used in b(ii) and (iii) indicates a numerical approach. If more than 1 mark is allocated it is essential to show your working, because any errors in the calculation will be marked consequentially. Incorrect answers may score some marks but only if you show your working. In part (c) of this question practical skills are tested within the written exam.

The marks allocated to each section indicate what is expected. For example, in part (a) there are 2 marks, so two observations are required. In (c)(ii) there are 4 marks, which indicates 1 mark for the diagram, 1 mark for the labels and 2 marks for identifying a possible error and suggesting a remedy for that error — 4 marks = four things wanted.

Student A

(a) Bubbles of carbon dioxide are given off.

Student B

(a) The white solid fizzes and a colourless solution is formed.

e Student A scores 1 mark but student B scores both marks. The clue is in the question. By looking carefully at the state symbols, you should be able to deduce what you would see. Student A has ignored the fact that 2 marks have been allocated so two observations are required.

Student A

(b) (i) 0.100 mol

■About this book

This guide is the first in a series of two, which together cover all of Modules 1, 2, 3 and 4 of the OCR AS Chemistry A (H032) and the OCR A Chemistry A (H432) specifications. It is written to help you prepare for examinations in either specification.

The **Content Guidance** section gives a point-by-point description of all the facts that you need to know and concepts that you need to understand for Modules 1 and 2. It aims to provide you with a basis for your revision. However, you must also be willing to use other sources in your preparation for the examination.

Module 1, Development of practical skills in chemistry, includes basic information that you will need for the whole of the A-level course. Practical skills, planning, implementation, analysis and evaluation are embedded throughout all the modules and will be assessed in written examinations.

Module 2, Foundations in chemistry, includes basic information that you will need for the whole of the A-level course. A good understanding of this module, particularly the mole concept and the use of balanced equations, will provide you with a solid foundation on which to build.

Modules 1 and **2** feature in all examinations whether AS or A-level.

The **Questions & Answers** section shows you the sort of questions that you can expect in the component test. It would be impossible to give examples of every kind of question in one book, but the questions used should give you a flavour of what to expect. Apart from question 22, which is for A-level students only, the questions are relevant to both AS and A-level students. Each question has been attempted by two students, Student A and Student B. Their answers, along with the comments, should help you to see what you need to do to score a good mark — and how you can easily *not* score marks, even though you probably understand the chemistry.

What can I assume about the guide?

You can assume that:

- the topics covered in the Content Guidance section relate directly to those in the specification
- the basic facts you need to know are stated clearly
- the major concepts you need to understand are explained
- the questions at the end of the guide are similar in style to those that will appear in the component tests
- the answers supplied are genuine, combining responses commonly written by students
- the standard of the marking is broadly equivalent to the standard that will be applied to your answers

What can I *not* assume about the guide?

You must *not* assume that:

- every last detail has been covered
- the way in which the concepts are explained is the *only* way in which they can be presented in an examination (often concepts are presented in an unfamiliar situation)
- the range of question types presented is exhaustive (examiners are always thinking of new ways to test a topic).

Study skills and revision techniques

All students need to develop good study skills. This section provides advice and guidance on how to study AS and first year A-level chemistry.

Organising your notes

Chemistry students often accumulate a large quantity of notes, so it is useful to keep these in a well-ordered and logical manner. It is necessary to review your notes regularly, maybe rewriting the notes taken during lessons so that they are clear and concise, with key points highlighted. You should check your notes using textbooks, and fill in any gaps. Make sure that you go back and ask your teacher if you are unsure about anything, especially if you find conflicting information in your class notes and textbook.

It is a good idea to file your notes in specification order using a consistent series of headings. The Content Guidance section can help you with this.

Organising your time

Preparation for examinations is personal. Different people prepare, equally successfully, in different ways. The key is being honest about what *works for you*.

Whatever your style, you must have a plan. Sitting down the night before the examination with your notes and a textbook does not constitute a revision plan — it is just desperation — and you must not expect a great deal from it. Whatever your personal style, there are some things you *must* do and some you *could* do.

The scheme outlined in the table opposite is a suggestion as to how you might revise Modules 1 and 2 over a 2-week period.

Day	Week 1	Week 2
	Each revision session should last approximately 30 min and during week 1 you should write brief summary notes on each topic.	
Mon	**30 min** on atoms, compounds, formulae and equations	Reread all your summary notes at least twice.
Tue	**20 min** on mole calculations **10 min** on atoms, compounds, formulae and equations	Module 1 refers to essential practical skills. Read the section in this book and make sure that you can do all the knowledge checks in this section. Allow about 30 min
Wed	**15 min** on acids and redox **10 min** on mole calculations **5 min** on atoms, compounds, formulae and equations	Module 1 refers to essential practical skills. Read the section in this book and make sure that you can do all the knowledge checks in this section. Select two questions from the back of the book. Cover up the exam advice comments and mark each student and then check to see if you agree with the comments. Allow about 30 min
Thurs	**15 min** on electron structure **10 min** on acids and redox **5 min** on mole calculations **2 min** on atoms, compounds, formulae and equations	Using the knowledge checks and questions at the end of this book or past papers or other question sources, try a structured question on a topic from Module 2. Mark it and list anything you do not understand. Allow about 30 min
Fri	**15 min** on bonding and structure **10 min** on electron structure **5 min** on acids and redox **2 min** on mole calculations **1 min** on atoms, compounds, formulae and equations	Using the knowledge checks and questions at the end of this book or past papers or other question sources, try a structured question on a topic from Module 2. Mark it and list anything you do not understand. Allow about 30 min
Sat	**15 min** on bonding and structure **10 min** on electron structure **5 min** on acids and redox **2 min** on mole calculations	Using the knowledge checks and questions at the end of this book or past papers or other question sources, try a structured question on a topic from Module 2. Mark it and list anything you do not understand. Allow about 30 min
Sun	**10 min** on bonding and structure **5 min** on electron structure **2 min** on acids and redox **You should now have gone through each topic at least three times and have short summary notes about each.**	Using the knowledge checks and questions at the end of this book or past papers or other question sources, try a structured question on a topic from Module 2. Mark it and list anything you do not understand Allow about 30 min

This revision timetable may not suit you, in which case write one to meet your needs. It is only there to give you an idea of how one might work. The most important thing is that the grid at least enables you to see what you should be doing and when you should be doing it. Do not try to be too ambitious — little and often is the best way.

It would of course be sensible to put together a longer rolling programme to cover all your AS or A-level subjects. Do *not* leave it too late. Start sooner rather than later.

Content Guidance

■ Module 1 Practical skills assessed in the written examination

Chemistry is a practical subject and the development of practical skills is essential. These skills will be assessed by your teachers. During the course you will be required to carry out a number of experiments, a minimum of 12, which are separately assessed. The experiments that you carry out will cover a range of technical skills and practical apparatus. Teachers will award a pass (or fail) to their students and performance in this component will be integral to, and examined in, all components of the 1 year and the full 2 year course.

The assessment will cover four key areas: planning, implementing, analysis and evaluation.

Table 1 gives a brief outline of the sorts of experiments you might be expected to encounter.

Year 1	Year 2
Mole determination	Rates of reaction
Acid–base titration	pH measurement
Qualitative testing of ions	Electrochemical cells
Preparation of an organic liquid	Redox titrations
	A range of organic syntheses and organic analysis
You will be expected to develop your research skills throughout the course.	

Table 1 Experiments you might encounter

Planning

Planning a scientific investigation is an essential skill and enables priorities to be dealt with in a controlled manner instead of simply reacting to things as they come along.

It is essential that you are able to identify the key stages in a scientific investigation and select the appropriate practical technique(s) and apparatus for a particular procedure. It is particularly important in any chemical experiment that you recognise that there are always significant safety risks that must be taken into account in the planning stages.

Once the hazards associated with a particular chemical have been identified, a **risk assessment** must be carried out. The risk assessment allows appropriate precautions to be put in place to allow the chemicals to be handled safely. Whenever carrying out a chemical experiment it is usual to wear a laboratory coat and possibly protective gloves but it is essential to *wear safety glasses at all times* when handling chemicals.

Knowledge check 1

Table 2 contains a number of hazard warning labels. Use the internet or look up a chemical catalogue to help you identify which hazard is represented by each warning label.

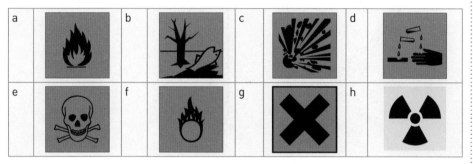

Table 2

Note: Some chemicals are too hazardous to be used in schools, therefore not all of these warning labels will be found in your chemistry laboratory.

Exam tip

It is impossible to remember individual hazards. You should always refer to a suitable database such as www.cleapss.org.uk/secondary/secondary-science/hazcards.

Identifying practical techniques and equipment

A crucial stage in planning a scientific investigation is to identify the most appropriate practical technique to allow you to safely carry out your experiment.

Knowledge check 2

Match the *number* of each laboratory technique in Table 3 to the *letter* of its most appropriate use.

	Technique		Use
1	Filtration	A	Weighing out chemicals
2	Water bath	B	Determining the concentration of a solution
3	Distillation	C	Heating an aqueous solution rapidly to 60°C
4	Balance	D	Separating a mixture of liquids
5	Bunsen burner	E	Collecting a gas that is insoluble in water
6	Gas syringe	F	Separating a solid from a liquid
7	Titration	G	Heating a flammable liquid to 120°C
8	Collecting a gas over water	H	Maintaining a reaction at 50°C
9	Heating mantle	I	Collecting a water-soluble gas

Table 3

Having identified the most appropriate technique for a particular scientific investigation, apparatus must then be selected to allow that technique to be carried out effectively and safely.

Knowledge check 3

Name the pieces of apparatus in Table 4 and state what each one is used for.

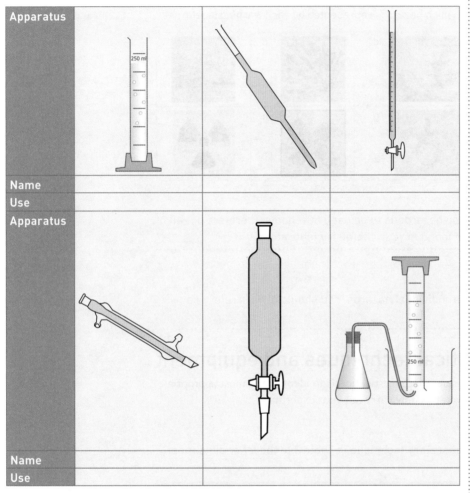

Apparatus		
Name		
Use		
Apparatus		
Name		
Use		

Table 4

Implementing

Each piece of laboratory apparatus used to measure a quantity has a limit to its precision. For example, a standard laboratory balance may give a measurement to the nearest 0.1 gram; greater precision may be achieved by using a balance capable of reading to three or four decimal places. For most measuring equipment, the manufacturer will give the maximum error that is inherent in using that piece of apparatus; this is sometimes etched onto the apparatus but, in other cases, will need to be looked up.

Examples of the tolerances/maximum errors of common laboratory equipment for measuring a single volume are given in the Table 5. Note that these values are examples only, the actual values will vary depending on the glassware available.

Apparatus	Tolerance (maximum error) /cm³
10 cm³ pipette	10 ± 0.04
25 cm³ pipette	25 ± 0.06
50 cm³ burette	50 ± 0.05
100 cm³ volumetric flask	100 ± 0.1
250 cm³ volumetric flask	250 ± 0.2
50 cm³ beaker	50 ± 5.0
100 cm³ beaker	100 ± 10.0
10 cm³ measuring cylinder	10 ± 0.5
100 cm³ measuring cylinder	100 ± 1.0

Table 5

The maximum error is an inevitable part of using that piece of equipment and it can be used for estimating the **percentage error**. A useful guide is that any apparatus has an error of one half of a graduation. If a 250 cm³ measuring cylinder has graduations every 5.0 cm³, the error is 2.5 cm³ for each measurement. If it is used to measure 120 cm³ of water, the percentage error is:

$$\frac{\text{error}}{\text{total}} \times 100 = \frac{2.5}{120} \times 100 = 2.1\%$$

An increased error arises if both the start and end graduations are used to measure a volume. For example, if 20 cm³ of water is measured by filling a 100 cm³ measuring cylinder to the 100 cm³ mark and then pouring off the water until the 80 cm³ mark is reached, both the start and the end graduations are subject to a maximum error of 0.5 cm³. It is possible that the starting volume could be 100.5 cm³ and the end volume 79.5 cm³. The possible error is therefore 1 cm³ in the 20 cm³ measured, which is 5%.

Volumetric equipment

This is covered fully in the section on titrations on pages 36–39. However, it is important to stress that when using volumetric apparatus the correct position of the meniscus is essential. Figure 1(a) shows the correct way to measure volume, taking account of the meniscus. All volumetric apparatus (pipettes, burettes and volumetric flasks) are manufactured such that the correct volume is obtained when the bottom of the meniscus sits exactly on the line.

Percentage error is (the maximum error/actual value) × 100

> **Exam tip**
>
> In an exam you will be told the maximum error that applies to a particular piece of apparatus.

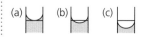

Figure 1

> **Exam tip**
>
> It is worth bearing in mind when planning an experimental procedure that it would probably be inappropriate to use a balance measuring to four decimal places (0.05% accuracy) and then to measure a volume with only 5% accuracy.

> **Knowledge check 4**
>
> a Calculate the percentage error in measuring 50 cm³ of a solution using:
> i a 250 cm³ measuring cylinder with a maximum error of 1 cm³
> ii a 50 cm³ pipette with a maximum error of 0.1 cm³
> iii a 50 cm³ burette where each reading has a maximum error of 0.05 cm³
> b Calculate the percentage error in measuring a mass of 2.5 g using a balance with a maximum error of   i 0.1 g   ii 0.05 g
> c Calculate the percentage error in measuring a temperature rise of 6°C using a thermometer with a maximum error of   i 0.5°C   ii 0.1°C

Recording results

When recording data, the precision should be indicated appropriately. For example, if you use a balance that reads to two decimal places, the masses recorded should indicate this. This may seem obvious for a mass of, for example, 2.48 g. However, you must remember that this applies equally for a mass of, for example, 2.50 g. Here the '0' should be included after the '5' to indicate that this mass is also precise to two decimal places. Recording the mass as 2.5 g is incorrect and could be penalised in an exam.

Burette readings are normally recorded to 0.05cm^3 as this represents the appropriate maximum error. So a reading of 23.45cm^3 is acceptable, but 23.47cm^3 is not. The reading from a burette should always be to two decimal places and always end with either .00 or .05. If the initial reading is zero, it should be written as 0.00cm^3.

Analysis

When carrying out experiments you will be expected to interpret both qualitative and quantitative data. In the first year of the course the qualitative analysis that you are expected to know is listed in section 3.1.4 of the specification, which details a range of chemical tests for ions, including: carbonate (CO_3^{2-}), sulfate (SO_4^{2-}), halides (Cl^-, Br^-, I^-) and ammonium (NH_4^+). The tests for each of these ions are fully covered in the student guide covering Modules 3 and 4 in this series.

You will encounter quantitative analysis when carrying out experiments involving moles (see pages 26–28) or enthalpy changes, which occur later in the course and are fully covered in the student guide covering Modules 3 and 4 in this series.

Significant figures

In simple cases the number of significant figures is simply the number of digits in the answer. 13.67 has 4 significant figures, while 13.7 has 3 significant figures and 13 has just 2.

In other cases numbers may need rounding up or down before quoting the answer to a particular number of significant figures. 71.64 has 4 significant figures but to 3 significant figures this is 71.6 (as 71.64 is nearer to 71.6 than to 71.7). To 2 significant figures it would be 72 (71.64 is nearer to 72 than to 71).

A number ending in a '5' is rather arbitrarily raised to the number above. So 11.5 must be written as 12 when quoted to 2 significant figures.

When a number such as 0.00754 has '0's *after* the decimal point these are not considered as 'significant', as shown in Figure 2. 0.00754 therefore has 3 significant figures.

When a number has '0's *before* the decimal point they are significant such that 1950.00 has 4 significant figures, as shown in Figure 3.

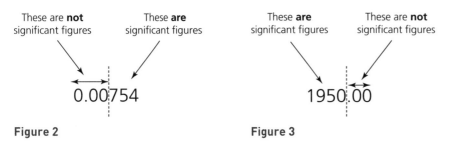

| These are **not** significant figures | These **are** significant figures | These **are** significant figures | These are **not** significant figures |

0.00754   1950.00

Figure 2   **Figure 3**

A number such as 2970 has 4 significant figures but if you are asked to quote this to 2 significant figures there is a temptation to quote this as '3000', which would be incorrect as 3000 has 4 significant figures. The way round it is to write the number in standard (index) form; 2970 written in standard form would be 3.0×10^3, which is to 2 significant figures.

Often your calculator will display an answer containing more digits than you were given in the data. Suppose you were asked to calculate the concentration of HCl(aq) in which $19.4\,cm^3$ of the HCl(aq) was neutralised by $25.0\,cm^3$ of $0.500\,mol\,dm^{-3}$ NaOH(aq). If you did this calculation correctly your calculator would show the concentration to be $0.644329896\,mol\,dm^{-3}$. The concentration of the solution isn't known to this degree of precision. The accuracy will be limited to the precision of the data or, in an experiment, by the accuracy of the apparatus. In the example above the data is given to 3 significant figures and the answer should also be limited to 3 significant figures. The figures after the third are dropped and the number is rounded. 0.644329896 when rounded to 3 significant figures is 0.644.

If the number after the significant number of figures is 4 or below the number is rounded down but if the number after the significant number of figures is 5 or above the number is rounded up.

4652.46 to 5 significant figures is   4652.5

4652.46 to 4 significant figures is   4652

0.0004352 to 3 significant figures is 0.000435

0.0004352 to 2 significant figures is 0.00044

When carrying out a calculation always quote your answer to the same number of significant figures given in the data. If the number of significant figures in the data varies, the *least* accurate should be used.

Knowledge check 5

Write the following numbers to 3 significant figures. You may have to use standard form.
a 734.8
b 69845.6
c 0.0003456
d Density = mass/volume. Calculate the density of a solution when:
   i 23 .1 g of solute was dissolved in $57\,cm^3$ of solution. Quote your answer to an appropriate number of significant figures.
   ii 12 .17 g of solute was dissolved in $30.0\,cm^3$ of solution. Quote your answer to an appropriate number of significant figures.
   iii 123 g of solute was dissolved in $1026.45\,cm^3$ of solution. Quote your answer to an appropriate number of significant figures.

Exam tip

When carrying out calculations it is essential that you do not round until the end of the calculation. If necessary you must use the 'memory' function on your calculator.

Using numbers in standard (index) form

Numbers can be written in different formats. A common way to write numbers is to use the decimal notation, for example 123642.78 and 0.0005432. However, when working with very large numbers (for example, 123642.78) or very small numbers (such as 0.0005432) it is convenient to write these numbers in standard notation.

Standard notation means writing the number as a product of two factors (Table 6):
- the first factor — the decimal point *always* comes after the first digit
- the second factor is *always* a multiple of 10

Number	Standard notation
4321	4.321×10^3
432.1	4.321×10^2
43.21	4.321×10^1
4.321	4.321×10^0 *
0.4321	4.321×10^{-1}
0.04321	4.321×10^{-2}
0.004321	4.321×10^{-3}
0.0004321	4.321×10^{-4}

* $10^0 = 1$, hence 4.321×10^0 is normally written as 4.321.

Table 6

One advantage of using standard numbers is that estimating the answer to a calculation is often quicker and easier.

Exam tip

It is always useful to estimate the answer to a calculation before doing the calculation on a calculator. It makes it easy to spot whether or not you have input the data into the calculator correctly.

Drawing graphs

Figure 4 shows a simple relationship between two variables.

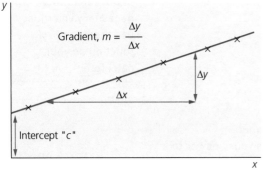

Figure 4 A simple relationship between two variables

The relationship between x and y is $y = mx + c$ (where m is the gradient and c is the intercept).

Δx is the change in x and Δy is the change in y.

Exam tip

The symbol Δ is used to represent 'change in ...', such that:
- ΔT is change in temperature
- ΔP is change in pressure
- ΔV is change in volume

When drawing graphs you should:

■ Choose a scale that will allow the graph to cover as much of the graph paper as possible. It is helpful to start both axes at zero. If all the points on one axis are between 90 and 100, to start at zero on that axis would cramp your graph into a small section of the paper (see Figure 5a). It is much better to truncate the axis so that the graph fills as much of the paper as possible (Figure 5b).

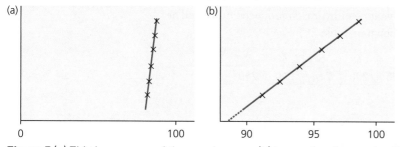

Figure 5 (a) This is poor use of the graph paper, (b) truncating the x-axis allows much better use of the graph paper

■ Label the axes with the dimensions and the units such as volume/cm³, and concentration/mol dm⁻³.

After plotting all the points on a graph, it may be that, due to experimental error, you may not get a perfect straight line or a curve that goes through all of the points. You will have to draw a line of best fit for the points (Figure 6).

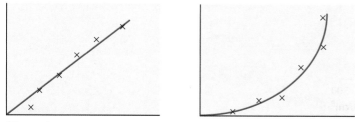

Figure 6

Drawing tangents to a curve

By calculating the gradient of the tangent (Figure 7) it is possible to work out how the concentration changes with respect to time after x seconds. This enables you to calculate the rate of reaction after x seconds. The units of rate are the units of y/x, which are mol dm⁻³/s, which is written as mol dm⁻³ s⁻¹.

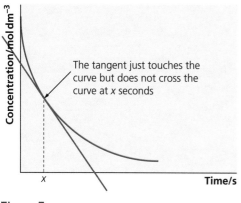

The tangent just touches the curve but does not cross the curve at x seconds

Figure 7

It is also important to ensure that you choose the axes appropriately and it is usual to plot the 'independent variable' on the *x*-axis and the 'dependent variable' on the *y*-axis.

Example

The reaction of dilute hydrochloric acid with 0.243 g magnesium ribbon was monitored over a period of time and the volume of hydrogen produced measured at fixed intervals of time. The results are shown in Table 7.

Time/min	1	2	3	4	5	6	7	8
Volume of $H_2(g)$/cm^3	78	146	205	220	234	240	240	240

Table 7

In this reaction the independent variable is the time. The volume of hydrogen depends on the time and is therefore the dependent variable.

If the axes are plotted incorrectly such that the independent variable is on the *y*-axis, as in Figure 8, the graph is difficult to interpret.

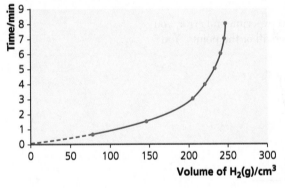

Figure 8

If, however, the independent variable is plotted on the *x*-axis, as in Figure 9, it is easy to interpret the graph and deduce that the reaction is:

■ quickest at the start of the reaction and gradually slows down
■ completed after 8 minutes
■ half completed after approximately 3 minutes

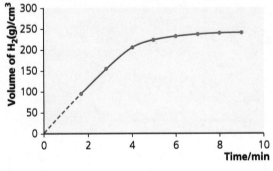

Figure 9

Evaluation

You should be able use your knowledge and understanding to evaluate your results and to draw valid conclusions. You should be able to identify anomalies in experiment data.

Knowledge check 6

A student carries out a titration and records his results (see Table 8). He wants to work out the average titre value. Which results in the table should he discard before calculating the average titre? Why should he discard these values? What is the average titre value?

Titration	Volume/cm^3
Rough	25.0
1	25.40
2	25.50
3	26.50

Table 8

Limitations in experimental procedure

Apart from the limitations imposed by apparatus, experiments also have errors caused by the procedure adopted. Such errors are difficult to quantify, but the following check list might help you to assess an experiment.

Handling substances

- Can you be sure that the substances you are using are pure?
- Does the experiment involve a reactant or a product that could react with the air?

Heating substances

- If you need to heat a substance until it decomposes, you can only be sure that the decomposition is complete if you heat to constant mass.
- Is it possible that the heating is too strong and the product has decomposed further?

Solutions

- If an experiment is quantitative, can you guarantee that any solution used is exactly at the stated concentration?

Gases

- If a gas is collected during the experiment, can you be sure that none has escaped?
- If you collect a gas over water, are you sure it is not soluble in water?
- The volume of a gas is temperature dependent.

Timing

- If an experiment involves timing, are you sure that you can start and stop the timing exactly when required?

Enthalpy experiments

- Heat loss is always a problem in enthalpy experiments, particularly if the reaction is slow.

Precision and accuracy of measurements and data

When performing a quantitative experiment it is likely that you will use several pieces of apparatus whose maximum errors will vary. Results should reflect the accuracy of each piece of apparatus used.

Knowledge check 7

A student performed a titration using a burette with a maximum error of $0.05\,cm^3$. The student's results are shown in Table 9 but there are a number of errors.

Titration	Rough	1	2	3
Final volume/cm³	24.0	23.57	23.9	23.62
Initial volume/cm³	0	0.1	0	0
Volume used/cm³	24	23.47	23.9	23.62

Table 9

a Copy Table 9 but give the results to the correct number of significant figures.

b State the mean value for the titre that should be used in subsequent calculations.

Improving experimental design

As with limitations resulting from procedure, it is not possible to provide a comprehensive list of ways to improve experiments. However, it is worth emphasising that simple changes are often effective.

Using more precise measuring equipment is only useful if it focuses on a significant weakness. For example, if the volume of a solution does not need to be measured precisely, there is no point in using volumetric equipment.

If a change in the apparatus seems necessary, look for a straightforward alternative. For example:

- If the problem is that a gas to be collected over water is slightly soluble, using a gas syringe might be an effective remedy.
- If a gas might escape from the apparatus when a reagent is added, the procedure could be improved by placing one reagent in a container inside the flask containing the other reagent and mixing them when the apparatus is sealed.
- If the problem with an enthalpy reaction is that it is too slow, using a powdered form would help.

It is a mistake to imagine that perfection can be achieved by using more complicated apparatus if the fault lies in the method that is being employed.

Summary

Having revised for the written exam of Module 1: **Development of practical skills in chemistry** you should now have an understanding of:
- planning: experimental design and the need to ensure that safe practices are adopted
- implementation: how to use practical apparatus and techniques correctly
- recording: observations and results appropriately
- analysis: interpreting qualitative and quantitative experimental data
- evaluation: how to draw conclusions, estimate errors and suggest valid improvements

Module 2 Foundations in chemistry

Atoms and reactions

Atomic structure and isotopes

Atoms consist of a nucleus containing protons and neutrons, which is surrounded by electrons (Table 10).

Particle	Relative mass	Relative charge	Distribution
Proton, p	1	+1	Nucleus
Neutron, n	1	0	Nucleus
Electron, e	1/1836	−1	Orbits/shells

Table 10

The **atomic number** and the **mass number** (see below) can be used to deduce the number of protons, neutrons and electrons in atoms and ions. Atoms are neutral and contain the same number of protons as electrons. Positive ions have lost electrons and hence have more protons than electrons, while negative ions have gained electrons, so they have fewer protons than electrons.

- The atomic number tells us the number of protons in an atom.
- The mass number tells us the number of protons plus the number of neutrons in an atom.
- The mass number minus the atomic number tells us the number of neutrons in an atom.

Atomic number, also known as the 'proton number', is the number of protons in the nucleus of an atom.

Mass number, also known as the 'nucleon number', is the number of protons and neutrons (nucleons) in the nucleus of an atom.

$$^{31}_{15}P \quad ^{32}_{15}P \quad ^{24}_{12}Mg^{2+} \quad ^{32}_{16}S^{2-}$$

^{31}P and ^{32}P are isotopes

15 p	15 p	12 p	16 p
15 e	15 e	10 e	18 e
16 n	17 n	12 n	16 n

Figure 10

As you can see from Figure 10, ^{31}P and ^{32}P are isotopes of phosphorus. **Isotopes** are atoms of the same element with different masses — they have the same number of protons (and electrons) but different numbers of neutrons. Chlorine has two isotopes — ^{35}Cl and ^{37}Cl.

Isotopes are atoms of the same element that have the same number of protons but different numbers of neutrons.

Relative mass

Most examinations ask for at least one or two definitions. Remember that 'define' is intended literally. Only a formal statement or equivalent paraphrase is required.

Definitions of relative isotopic mass and relative atomic mass are often required, and each one is based on the carbon-12 scale.

A similar definition is needed for each:

- The **relative isotopic mass** is the mass of an atom/isotope of the element compared with one-twelfth of the mass of a carbon-12 atom.
- The **relative atomic mass** of an element is the weighted mean mass of an atom of the element compared with one-twelfth of the mass of a carbon-12 atom.

It is possible to use this basic definition and amend it slightly to create definitions of the relative molecular mass and the relative formula mass:

- The **relative molecular mass** is the weighted mean mass of a molecule compared with one-twelfth of the mass of a carbon-12 atom.
- The **relative formula mass** is the weighted mean mass of a formula unit compared with one-twelfth of the mass of a carbon-12 atom.
- Both definitions are required as relative molecular mass only refers to covalent molecules whereas relative formula mass applies to all compounds, including ionic substances.

Relative molecular mass and relative formula mass can be calculated by using the relative atomic masses.

Example 1

Calculate the relative molecular mass of (a) carbon dioxide, CO_2 and (b) glucose, $C_6H_{12}O_6$.

Answer

(a) $CO_2 = 12 + 16 + 16 = 44$

(b) $C_6H_{12}O_6 = (12 \times 6) + (1 \times 12) + (16 \times 6)$

$= 72 + 12 + 96 = 180$

Example 2

Calculate the relative formula mass of sodium carbonate, Na_2CO_3.

Answer

$Na_2CO_3 = (23 \times 2) + 12 + (16 \times 3)$

$= 46 + 12 + 48 = 106$

Example 3

Calculate the relative formula mass of copper sulfate crystals, $CuSO_4.5H_2O$.

Answer

$CuSO_4.5H_2O = 63.5 + 32.1 + (16 \times 4) + 5 \times (1 + 1 + 16)$

$= 63.5 + 32.1 + 64 + 90 = 249.6$

Exam tip

Students often lose marks by careless use of words, for instance sodium chloride, NaCl, must *not* be described as a molecule — it is a giant lattice made up of millions of ions and does not exist as a single NaCl molecule.

Knowledge check 8

Define relative molecular mass, atomic number and mass number.

Exam tip

Students often make mistakes when working out relative formula mass of hydrated crystals like $CuSO_4.5H_2O$. The correct answer is 249.6, as shown in example 3, but lots of students give the answer incorrectly as 185.6 by misreading the $5H_2O$ in the formula.

Knowledge check 9

Calculate the relative formula mass of $(NH_4)_3PO_4$ and $CH_3C_6H_2(NO_2)_3$.

You will be expected to calculate the relative atomic mass from data, where you will be given information about each isotope and told the relative abundance of each isotope.

Example 1

A sample of chlorine contains about 75% ^{35}Cl and 25% ^{37}Cl. Calculate the relative atomic mass of chlorine.

Answer

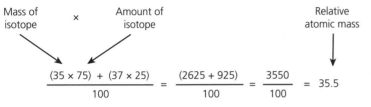

$$\frac{(35 \times 75) + (37 \times 25)}{100} = \frac{(2625 + 925)}{100} = \frac{3550}{100} = 35.5$$

Figure 11

Example 2

The mass spectrum of a sample of lithium is shown in Figure 12. Use it to calculate the relative molecular mass of lithium.

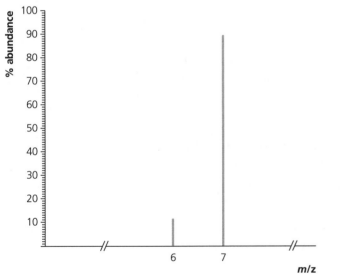

Figure 12 Mass spectrum of lithium

Answer

The percentage of each isotope is 11% ^6Li and 89% ^7Li. The relative formula mass is:

$$\frac{(6 \times 11) + (7 \times 89)}{100} = \frac{66 + 623}{100} = \frac{689}{100} = 6.89$$

HARROW COLLEGE
Learning Centre

Exam tip

The first question on the paper is usually designed to be straightforward and is often about isotopes and the calculation of relative atomic mass of an element using data from mass spectrometry. Examples 1 and 2 show the calculation that often forms part of a typical opening question.

Knowledge check 10

A sample of iron contains three isotopes: ^{54}Fe, ^{56}Fe and ^{57}Fe. The relative abundance is 2:42:1, respectively. Calculate the relative atomic mass of Fe.

Exam tip

All of the written exams will contain 'synoptic' questions that will be used to test your ability to apply your knowledge and understanding. A question like example 3 could be used to meet these requirements.

Knowledge check 11

Use this mass spectrum of magnesium to calculate the relative atomic mass of magnesium.

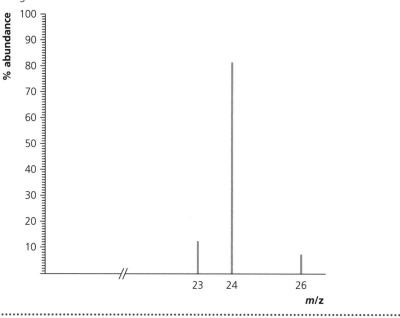

Example 3

A sample of boron was known to contain two different isotopes, $^{10}_{5}B$ and $^{11}_{5}B$. The relative atomic mass of the sample of boron was 10.8. Calculate the percentage of each isotope in the sample.

Answer

This is a more difficult calculation, which requires the application of some basic maths. The standard equation for calculating the relative atomic mass is:

$$\frac{(\text{mass of isotope 1} \times \% \text{ of isotope 1}) + (\text{mass of isotope 2} \times \% \text{ of isotope 2})}{100}$$

The two isotopes account for 100% of the boron in the sample so if there is $x\%$ of ^{10}B there must be $(100 - x)\%$ of ^{11}B and together they have a relative atomic mass of 10.8. The equation can be written as:

$$\frac{(10x) + [11(100 - x)]}{100} = 10.8$$

Therefore $10x + 1100 - 11x = 1080$, such that

$$-x = 1080 - 1100$$

$$= -20$$

$$\text{so } x = 20\%$$

The sample of boron contains 20% ^{10}B and 80% ^{11}B

Compounds, formulae and equations

Formulae and equations

It is essential that you can write the formulae of a range of common chemicals and write balanced equations. The chemistry of group 2 and group 17 elements (see the student guide covering Modules 3 and 4 in this series) provides many opportunities to practise this. You should also look back to your GCSE notes and practise writing formulae and equations.

The periodic table can be used to deduce the formula of most chemicals, although there are many exceptions.

Group	1	2	13	14	15	16	17
Number of bonds (valency)	1	2	3	4	3	2	1

Table 11

If a group 1 element forms a compound with a group 16 element:
- the group 1 element, for example, Li, forms one bond
- the group 16 element, for example, O, forms two bonds
- two lithiums are required for each oxygen
- therefore the formula of the compound is Li_2O

If a group 13 element forms a compound with a group 16 element:
- the group 13 element, for example, Al, forms three bonds
- the group 16 element, for example, O, forms two bonds
- two aluminiums are required for three oxygens (each equates to six bonds)
- therefore the formula of the compound is Al_2O_3

An alternative way of deducing a formula is to use the 'valency cross-over' technique. Using aluminium oxide as an example, follow the simple steps:

Step 1   Write each of the symbols   Al   O

Step 2   Write the valency on the top right-hand side of the symbol Al^3 O^2

Step 3   Cross-over the valencies and write them at the bottom
right of the other symbol $Al_2^3 \diagdown O_3^2$

Step 4   You now have the formula of the compound, Al_2O_3

You are also expected to know the formulae of hydrochloric acid (HCl), sulfuric acid (H_2SO_4), nitric acid (HNO_3) and their corresponding salts.

Valency		
1	All group 1 elements, H, and Ag	All group 17 elements, OH (hydroxide) NO_3 (nitrate), HCO_3 (hydrogencarbonate) and NH_4 (ammonium)
2	All group 2 elements, Fe, Cu, Zn, Pb, Sn	O, S, SO_4 (sulfate), CO_3 (carbonate)
3	All group 13 elements and Fe	
4	C, Si, Pb, Sn	

Table 12 Valencies of elements and groups of elements

Exam tip

It is important that you learn all of these valencies — if you get a formula wrong it usually means you will get the equation wrong, as well as any calculations that follow. Getting a formula right ensures that you won't lose more marks in a calculation that follows.

You should be able to use Table 12 to work out most formulae. You will also be expected to learn and recall the formula of various ions, including those listed in Table 13.

Positive ions (cations)	Negative ions (anions)
Ammonium, NH_4^+	Nitrate, NO_3^-
Zinc, Zn^{2+}	Carbonate, CO_3^{2-}
Silver, Ag^+	Sulfate, SO_4^{2-}
Iron(II), Fe^{2+}	Hydroxide, OH^-
Iron(III), Fe^{3+}	

Table 13 Common cations and anions

More about equations

Being able to provide the correct formulae for substances is important because this enables you to write equations for chemical reactions. An equation not only summarises the reactants used and the products obtained, it also indicates the numbers of particles of each substance that are required or produced.

A simple case is the reaction of carbon and oxygen to make carbon dioxide. This is summarised in an equation as:

$$C + O_2 \rightarrow CO_2$$

The equation tells us that one atom of carbon reacts with one molecule of oxygen to make one molecule of carbon dioxide.

When hydrogen reacts with oxygen:

$$2H_2 + O_2 \rightarrow 2H_2O$$

we see that two molecules of hydrogen are required for each molecule of oxygen to make two molecules of water.

In any equation all symbols must be balanced and there must be the same number of each on both sides of the equation. You may be asked to include state symbols, such that the equations for the reaction of oxygen with either carbon or with hydrogen would be written:

$$C(s) + O_2(g) \rightarrow CO_2(g) \text{ and } 2H_2(g) + O_2(g) \rightarrow 2H_2O(l)$$

Amount of substance

The mole

The **mole** is the SI unit for the amount of substance that contains as many single particles as there are atoms in 12 g of the carbon-12 (^{12}C) isotope, and is equal to **Avogadro's constant**, N_A, which has a value of 6.02×10^{23} mol^{-1}.

The mass of 1 mole of molecules of a substance equals the relative molecular mass in grams. This is the **molar mass**, M, and it is defined as the mass, in grams, per mole of a substance. The units of molar mass are g mol^{-1}. The amount of substance in moles is given the symbol 'n'.

> **Knowledge check 12**
>
> What are the formulae of sodium oxide, calcium hydroxide, ammonium sulfate, aluminium nitrate and zinc carbonate?

> 1 **mole** of any substance is the amount of substance (in grams) that contains as many specified entities (atoms, molecules, ions etc.) as there are in 12 g of the carbon-12 isotope.

Determination of formulae

It is important to understand the difference between **empirical** and **molecular formulae**.

The **empirical formula** is defined as the simplest whole number ratio of atoms of each element in a compound. The **molecular formula** is defined as the actual number of atoms of each element in a molecule of a compound.

Example

Compound A has a relative molecular mass of 180 and composition by mass of C, 40%; H, 6.7%; O, 53.3%. Calculate the empirical formula and the molecular formula.

Answer

Divide the percentage of each element by its own relative atomic mass:

$$\text{C} \qquad\qquad \text{H} \qquad\qquad \text{O}$$

$$\frac{40}{12} = 3.3 \qquad \frac{6.7}{1} = 6.7 \qquad \frac{53.3}{16} = 3.33$$

Divide each by the smallest:

$$\frac{3.3}{3.3} = 1 \qquad \frac{6.7}{3.33} = 2 \qquad \frac{3.33}{3.33} = 1$$

Hence the empirical formula is $C_1H_2O_1 = CH_2O$

Calculate the mass of the empirical formula:

$$CH_2O = 12 + 2 + 16 = 30$$

Deduce how many empirical units are needed to make up the molecular mass:

$$\frac{\text{molecular mass}}{\text{empirical mass}} = \frac{180}{30} = 6$$

Therefore the molecular formula is made up of six empirical units, so the molecular formula is $C_6H_{12}O_6$

Exam tip

When carrying out a calculation, keep the numbers in your calculator and only 'round' at the end of the calculation.

Knowledge check 13

Deduce the molecular formula of a hydrocarbon that contains 90.6% C and has a molar mass of $106\,\text{g}\,\text{mol}^{-1}$.

Anhydrous salts, hydrated salts and water of crystallisation

Salts such as copper sulfate can exist as either the anhydrous salt, $CuSO_4$, or as the hydrated salt, $CuSO_4.5H_2O$. Anhydrous copper sulfate is a white powder, while hydrated copper sulfate exists as shiny blue crystals. They appear crystalline due to the water of crystallisation.

You are expected to be able to calculate the formula of a hydrated salt from percentage composition data or from experimental data.

Example

2.86 g of hydrated sodium carbonate, $Na_2CO_3.\,xH_2O$ was heated to constant mass. The mass of the remaining anhydrous sodium carbonate, Na_2CO_3, was 1.06 g. Use these data to calculate the value of x.

Answer

mass of water in $Na_2CO_3.xH_2O = 2.86 - 1.06 = 1.80\,\text{g}$

molar mass of water $= 1 + 1 + 16 = 18$

$\rightarrow$

$$\text{moles of water} = \frac{1.80}{18} = 0.1\,\text{mol}$$

mass of anhydrous $Na_2CO_3 = 1.06\,\text{g}$

molar mass of $Na_2CO_3 = 23 + 23 + 12 + (3 \times 16) = 106$

$$\text{moles of anhydrous } Na_2CO_3 = \frac{1.06}{106} = 0.01\,\text{mol}$$

	Na_2CO_3		H_2O
moles	$\dfrac{0.01}{0.01}$	:	$\dfrac{0.1}{0.01}$
=	1	:	10

The formula of the hydrated crystals is $Na_2CO_3.10H_2O$, such that the value of x is 10.

Calculation of reacting masses, gas volumes and mole concentrations

Moles from mass

The amount of substance in moles (n) = the mass of substance (m) in g, divided by the molar mass of substance, M, such that $n = m/M$.

This applies to atoms, to covalent molecules and to ionic substances. The examples below illustrate the sort of questions that you may be asked. The triangle in Figure 13 can be used to help you rearrange the equation.

Example 1

Calculate the number of moles present in 8.0 g of NaOH.

Answer

The information given includes the mass, m (8.0 g), and the formula (NaOH), so it is possible to calculate the molar mass, M, of NaOH by adding together the relative atomic masses of each individual element in the molecule.

Na = 23; O = 16; H = 1

Therefore $M = 23 + 16 + 1 = 40$

$$n = \frac{m}{M}$$

Therefore $n = \dfrac{8.0}{40} = 0.2\,\text{mol}$

Knowledge check 14

Calculate the percentage, by mass, of water in citric acid, $HOC(CH_2)_2(COOH)_3.H_2O$.

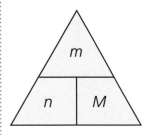

Figure 13

Example 2

Calculate the mass of 0.25 mol of $CaCO_3$.

Answer

The information given includes n, the amount in moles = 0.25 mol

You can use the formula ($CaCO_3$) to deduce the molar mass of $CaCO_3$. M is calculated by adding together the relative atomic masses of each individual element in the molecule.

Ca = 40.1; C = 12; O = 16

but there are three O atoms, therefore $3 \times 16 = 48$, so $M = 100.1$

$$n = \frac{m}{M}, \text{ which can be rearranged to give } nM = m$$

$$0.25 \times 100.1 = m$$

Therefore m is 25.025 g

It would be unusual to quote an answer to five significant figures, and you may be asked to quote the answer to three or four significant figures, in which case m is 25.0 g (to three significant figures), or 25.03 g (to four significant figures).

Mole calculations for gases

Chemical reactions also take place in the gas phase, and it is therefore important to be able to calculate amounts of chemicals present in gases. Avogadro deduced that all gases, under the same conditions of temperature and pressure, occupy the same volume. At room temperature and pressure the volume of one mole of a gas is equal to 24 dm³ or 24 000 cm³. This is known as the **molar gas volume** and has units $dm^3 \, mol^{-1}$ (or $cm^3 \, mol^{-1}$).

It follows that the number of moles of a gas n can be calculated using:

■ volume of gas in dm³:

$$n = \frac{V \text{ (in dm}^3)}{24 \text{ (in dm}^3 \text{mol}^{-1})}$$

■ volume of gas in cm³:

$$n = \frac{V \text{ (in cm}^3)}{24\,000 \text{ (in cm}^3 \text{mol}^{-1})}$$

See Figure 14.

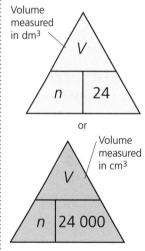

Knowledge check 15

a 0.05 mol of compound X weighs 9.0 g. Calculate the molar mass of compound X.

b Calculate the amount in moles in 14.30 g $Na_2CO_3.10H_2O$

Volume measured in dm³

V

n | 24

Volume measured in cm³

V

n | 24 000

Figure 14

Example

Calculate the number of moles present in $120\,cm^3$ of hydrogen at room temperature and pressure.

Answer

Use $n = \dfrac{V}{24\,000}$ (since the volume of gas has been given in cm^3):

$$n = \frac{120}{24\,000} = 0.005 = 5 \times 10^{-3}\,\text{moles}$$

Mole calculations for solutions

Many chemical reactions are carried out in solution (usually dissolved in water). The amount of chemical present is best described by using the concentration of the solution, c, in $mol\,dm^{-3}$, and the volume of the solution, V, in dm^3.

The amount in moles for solutions can be calculated using:

$$n = c \times V$$

(see Figure 15).

If V is given in cm^3, then use:

$$n = \frac{c \times V\,(\text{in }cm^3)}{1000}$$

The examples illustrate the sort of questions that you may be asked.

Example 1

Calculate the number of moles present in $250\,cm^3$ of $0.50\,mol\,dm^{-3}$ solution.

Answer

The concentration, c, is given as $0.50\,mol\,dm^{-3}$, and the volume, V, is $250\,cm^3$. The volume must be converted into dm^3, that is, $V = \dfrac{250}{1000} = 0.25\,dm^3$.

$$n = c \times V = 0.5 \times 0.25 = 0.125\,mol$$

Example 2

Calculate the concentration of a sodium hydroxide (NaOH) solution when $4.0\,g$ NaOH is dissolved in $250\,cm^3$ of water.

Answer

This is slightly more complicated, and you first have to calculate the number of moles of NaOH using the mass ($4.0\,g$) and the formula, NaOH, to deduce the molar mass ($23 + 16 + 1 = 40$), such that:

$$n = \frac{m}{M} = \frac{4}{40} = 0.1\,mol$$

The concentration can then be calculated, using:

$$c = \frac{n}{V} = \frac{0.1}{\left(\dfrac{250}{1000}\right)} = \frac{0.1}{0.25} = 0.4\,mol\,dm^{-3}$$

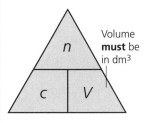

At room temperature and pressure, calculate the volume, in cm^3, of:

a $0.05\,mol\,SO_2$

b $1.28\,g\,O_2(g)$

Volume **must** be in dm^3

Figure 15

Calculate the concentration of a solution made by dissolving:

a $0.2\,mol\,NaCl$ in $500\,cm^3$ solution

b $2.496\,g\,CuSO_4.5H_2O$ in $250\,cm^3$ solution

Ideal gas equation

There is a significant difference between gases and solids or liquids in that the particles of a gas are very much further apart than the particles in a solid or a liquid. The volume of a gas depends on the temperature, the pressure and the number of moles of the gas.

- If the temperature, T, of a gas is increased, the volume will increase. This can be written as:

$$V \propto T$$

- If the pressure, P, of a gas is increased, the volume will decrease. This can be written as:

$$V \propto \frac{1}{P}$$

- If the amount in moles, n, of a gas is increased, the volume will increase. This can be written as:

$$V \propto n$$

By combining the three equations we obtain:

$$V \propto \frac{nT}{P}$$

which can be rearranged to give:

$$PV \propto nT$$

It follows that:

$$\frac{P_1V_1}{T_1} = \frac{P_2V_2}{T_2}$$

where P_1, V_1 and T_1 are the initial pressure, volume and temperature and P_2, V_2 and T_2 are the final pressure, volume and temperature.

Adding a constant of proportionality, R, results in the relationship $PV = nRT$, which is known as the **ideal gas equation**. The constant of proportionality is known as **the gas constant** and has the value $8.314\,\mathrm{J\,K^{-1}\,mol^{-1}}$ (joules per kelvin per mole).

As the name suggests, the ideal gas equation only applies to an **ideal gas**.

Real gases deviate from ideal gases but do approach ideal behaviour at very high temperatures and at very low pressures.

Example 1

If the volume of a gas collected at 50°C and 110 kPa is 75 cm³, what would the volume be at standard temperature and pressure (STP), where temperature is 273 K (0°C) and pressure is 101 kPa?

Answer

For calculations of this type it is best to use the equation:

$$\frac{P_1V_1}{T_1} = \frac{P_2V_2}{T_2}$$

The only unknown is V_2, so the equation can be rearranged to give:

→

Exam tip

The value and units of the gas constant will be on the data sheet supplied in all examinations.

An **ideal gas** is composed of independent particles (molecules or atoms), which are widely separated from each other. An ideal gas can be described by the following:

- the molecules (or atoms) are in continuous random motion
- there are no intermolecular forces between the individual molecules (or atoms)
- all collisions are perfectly elastic with no exchange of kinetic energy
- the molecules (or atoms) have no size (that is, they occupy zero volume)

$$\frac{P_1 V_1 T_2}{P_2 T_1} = V_2$$

Pressure is in kPa. Standard pressure is 101 kPa.

Temperatures must be in K, so if $T_1 = 50°C$, then it is $50 + 273 = 323\,K$.

Standard temperature is $0°C = 273\,K$.

The units of the calculated volume V_2 will be the same as the units used for V_1.

$$\frac{110 \times 75 \times 273}{101 \times 323} = \frac{22\,522\,500}{32\,623} = 69\,cm^3$$

Example 2

2.63 g of a noble gas occupies a volume of $500\,cm^3$ at RTP (room temperature and pressure, where room temperature is 298 K and pressure is 101 kPa). Identify the noble gas.

Answer

Use the ideal gas equation, $PV = nRT$. Since we are using RTP, the only unknown is the amount in moles, n.

The ideal gas equation can be rearranged to give

$$n = \frac{PV}{RT}$$

$$n = \frac{101 \times 0.5}{8.314 \times 298} = \frac{50.5}{2477.6} = 0.02\,mol$$

We now know that 0.02 mol of the noble gas has a mass = 2.63 g.

Therefore, the relative atomic mass of the noble gas can be written as:

$$\text{relative atomic mass} = \frac{m}{n} = \frac{2.63}{0.02} = 131.3$$

Therefore the noble gas is xenon (xenon has an atomic mass = $131.3\,g\,mol^{-1}$).

Example 3

At RTP, 0.76 g of a halogen gas occupies a volume of $490\,cm^3$. Identify the halogen gas.

Answer

Start with the equation $PV = nRT$

$$n = \frac{PV}{RT} = \frac{101 \times 0.490}{8.314 \times 298} = 0.0199\,mol$$

$$\text{molar mass} = \frac{0.76}{0.0199} = 38.0\,g\,mol^{-1}$$

The halogen gas is fluorine, F_2, which has a molar mass of $2 \times 19.0 = 38.0\,g\,mol^{-1}$.

Using equations

Mole calculations appear on almost every exam paper. You should be able to:

- write and balance full (and ionic) equations
- calculate reacting masses
- calculate reacting gas volumes
- calculate reacting volumes of solutions
- calculate concentrations from titrations

The calculation in the next example shows how reacting masses, reacting gas volumes and reacting volumes of solutions could all be tested in a single question.

Example

Zinc reacts with dilute hydrochloric acid to produce zinc chloride and hydrogen gas. Write a balanced equation for this reaction and calculate the mass of zinc required to react exactly with $50\,cm^3$ of a $0.10\,mol\,dm^{-3}$ solution of HCl(aq). Deduce the volume, in cm^3, of hydrogen gas produced in this reaction.

Answer

Equation: $Zn + 2HCl \rightarrow ZnCl_2 + H_2$

The number in front of each formula in the balanced equation tells us the number of moles used and gives us the ratio of the reacting moles, i.e. the mole ratio:

$$Zn \quad + \quad 2HCl \quad \rightarrow \quad ZnCl_2 \quad + \quad H_2$$

mole ratio: 1 mol : 2 mol : 1 mol : 1 mol

We can calculate n for HCl, since we are given c and V.

$$n = \frac{0.1 \times 50}{1000} = 0.0005 = 5 \times 10^{-3}\,mol$$

Once we know the moles of one chemical in the equation, we can deduce the number of moles of all the other chemicals in the equation by using the mole ratio.

$$Zn \quad + \quad 2HCl \quad \rightarrow \quad ZnCl_2 \quad + \quad H_2$$

actual moles: $\dfrac{5 \times 10^{-3}}{2}$ $\quad 5 \times 10^{-3} \quad$ $\dfrac{5 \times 10^{-3}}{2}$ $\dfrac{5 \times 10^{-3}}{2}$

$\quad\quad 2.5 \times 10^{-3} \quad\quad\quad\quad\quad\quad\quad 2.5 \times 10^{-3} \quad 2.5 \times 10^{-3}$

We can use $n = \dfrac{m}{M}$ to find the mass of Zn required. We can now use $n = \dfrac{V}{24\,000}$ to find the volume of H_2 produced.

Zn	H_2
$n = \dfrac{m}{M}$	$n = \dfrac{V}{24\,000}$
$2.5 \times 10^{-3} = \dfrac{m}{65.4}$	$2.5 \times 10^{-3} = \dfrac{V}{24\,000}$
$2.5 \times 10^{-3} \times 65.4 = m$	$2.5 \times 10^{-3} \times 24\,000 = V$
$0.16 = m$	$60 = V$
Mass of Zn required = $0.16\,g$	Volume of H_2 produced = $60\,cm^3$

Exam tip

Every exam will test your understanding of the mole. Questions often link together reactions that require you to use reacting mole ratios as well as combining the calculation of the number of moles for solids, gases and solutions.

Knowledge check 18

For the reaction:

$$MgCO_3(s) + 2HCl(aq) \rightarrow MgCl_2(aq) + H_2O(l) + CO_2(g)$$

state what you would see in the reaction and calculate:

a the volume of $0.05\,mol\,dm^3$ HCl(aq) required to react with $0.843\,g$ $MgCO_3$(s)
b the volume of CO_2(g) that would be produced

Knowledge check 19

Excess zinc powder was added to a $50\,cm^3$ of $0.01\,mol\,dm^{-3}$ solution of $CuSO_4$(aq).

$$Zn(s) + CuSO_4(aq) \rightarrow ZnSO_4(aq) + Cu(s)$$

Calculate the mass of Cu(s) precipitated.

Percentage yields and atom economy

Percentage yield calculations

Percentage yield calculations involve mole calculations and are often used when preparing organic compounds. Reactions of organic molecules will be covered fully in module 4 and module 6. A typical calculation is shown here:

$$\text{percentage yield} = \frac{\text{actual yield}}{\text{theoretical yield}} \times 100$$

Example

Ethanol, C_2H_5OH, can be oxidised to form ethanoic acid, CH_3COOH. If $2.3\,g$ of ethanol are oxidised to produce $2.4\,g$ of ethanoic acid, calculate the percentage yield.

Answer

Any mole calculations require a balanced equation, so it is essential that you are able to write suitable balanced equations and to use the mole ratios from the equation. [O] can be used to represent the oxidising agent.

Equation: C_2H_5OH +   2[O] $\rightarrow$ CH_3COOH + H_2O

Mole ratio: 1 mole : 2 moles : 1 mole : 1 mole

The equation shows that 1 mole of ethanol produces 1 mole of ethanoic acid.

Step 1: calculate the number of moles of ethanol used:

$$n = \frac{m}{M}$$

amount of ethanol used $= n = \dfrac{\text{mass (in g)}}{\text{molar mass}}$

$$= \frac{2.3}{46} = 0.05 \text{ moles}$$

Since the mole ratio is 1 : 1, the amount of ethanoic acid that could be made is also 0.05 moles.

Theoretical yield is the mass of the product assuming that the reaction goes according to the chemical equation and the reaction is 100% efficient.

Step 2: Calculate the number of moles of ethanoic acid actually produced:

$$\text{amount of ethanoic acid produced} = n = \frac{m}{M}$$

$$= \frac{2.4}{60} = 0.04 \text{ moles}$$

Step 3: calculate the percentage yield by using:

$$\text{percentage yield} = \frac{\text{actual yield}}{\text{maximum yield}} \times 100$$

$$= \frac{0.04 \times 100}{0.05} = 80\%$$

Knowledge check 20

A student reacted 5.00 g of ethanol (C_2H_5OH) with 8.00 g of ethanoic acid (CH_3COOH) and made 7.12 g of ethyl ethanoate ($CH_3COOC_2H_5$). Calculate the student's percentage yield. The equation for the reaction is:

$$C_2H_5OH + CH_3COOH \rightleftharpoons CH_3COOC_2H_5 + H_2O$$

Atom economy calculations

$$\text{atom economy} = \left(\frac{\text{molar mass of the desired products}}{\text{total molar mass of all the products}}\right) \times 100$$

The reaction between ethanol, C_2H_5OH, and ethanoic acid, CH_3COOH, is used to make the ester ethyl ethanoate, $CH_3COOC_2H_5$.

$$CH_3COOH + C_2H_5OH \rightarrow CH_3COOC_2H_5 + H_2O$$

The desired product is the ester but water is also produced.

molar mass of $CH_3COOC_2H_5 = 88\,g$

molar mass of $H_2O = 18\,g$

$$\text{atom economy} = \left(\frac{88}{88 + 18}\right) \times 100$$

$$= \left(\frac{88}{106}\right) \times 100 = 83\%$$

> **Exam tip**
>
> If asked to define atom economy, simply use this equation.

A high atom economy is good, since that would indicate a low level of waste. Although concern over atom economy is a relatively new idea, its importance is likely to grow as society becomes ever more concerned about the need to conserve resources and reduce the production of unwanted by-products. Atom economy can be greatly enhanced if a use can be found for by-products.

> **Exam tip**
>
> Do not round numbers during the calculation. Keep the number in your calculator and only round when you have finished the entire calculation.

Knowledge check 21

A student prepared a sample of ethanol, C_2H_5OH, by reacting bromoethane with sodium hydroxide:

$$C_2H_5Br + NaOH \rightarrow C_2H_5OH + NaBr$$

Calculate the atom economy for the preparation of ethanol.

Acids

Acids and bases

Acids are defined as proton donors, and bases (alkalis) can be described as proton acceptors.

The proton, H^+, released by an acid can only occur in aqueous solution. Pure hydrogen chloride, for example, is a covalent gas, and it is only when it comes into contact with water that it can release a proton.

$$HCl(g) + H_2O(l) \rightarrow H_3O^+(aq) + Cl^-(aq)$$

$H_3O^+(aq)$ is often written as $H^+(aq)$, and the equation is often given as:

$$HCl(aq) \rightarrow H^+(aq) + Cl^-(aq)$$

You are expected to know the formulae of some common acids and alkalis. These are listed in Table 14, and you will *not* be given these formulae in an examination.

Common acids	Common alkalis
Hydrochloric acid, HCl	Sodium hydroxide, NaOH
Sulfuric acid, H_2SO_4	Potassium hydroxide, KOH
Nitric acid, HNO_3	Ammonia, NH_3
Ethanoic acid, CH_3COOH	Ammonium hydroxide, NH_4OH

Table 14 Common acids and alkalis

Acids have a pH below 7, and the stronger acids have lower pHs. Acids can be subdivided into strong acids and weak acids.

- Strong acids include HCl, H_2SO_4 and HNO_3. They are regarded as strong acids because in solution they totally dissociate into their ions. The equation always includes the '$\rightarrow$' symbol.

$$HCl(aq) \rightarrow H^+(aq) + Cl^-(aq)$$

- Weak acids include organic acids such as ethanoic acid, CH_3COOH. They are weak acids because in solution they only partially dissociate into their ions. The equation always includes the '$\rightleftharpoons$' symbol.

$$CH_3COOH(aq) \rightleftharpoons CH_3COO^-(aq) + H^+(aq)$$

Bases are defined as proton (H^+) acceptors. An alkali is defined as a soluble base that can release hydroxide ions, OH^-, when in aqueous solution. $NaOH(aq)$ and $KOH(aq)$ are strong alkalis, while $NH_3(aq)$ and $NH_4OH(aq)$ are weak alkalis.

Common bases include metal oxides (for example, Na_2O, MgO, CuO), metal hydroxides (for example, NaOH, KOH, $Cu(OH)_2$) and ammonium hydroxide (NH_4OH) and metal carbonates (Na_2CO_3, $ZnCO_3$).

Common alkalis include group 1 hydroxides such as LiOH and NaOH and some group 2 hydroxides, such as $Ca(OH)_2$.

> **Exam tip**
>
> Make sure you know the formulae of the common acids, HCl, H_2SO_4, HNO_3 and CH_3COOH, and that you are able to work out the formulae of the salts they form.

Neutralisation

Neutralisation occurs when an acid reacts with a base to form a salt and water. Neutralisation reactions occur when an acid reacts with an alkali, a metal oxide or a carbonate.

Salts

A salt is defined as a substance that is formed when one or more H^+ ions of an acid are replaced by a metal ion or an ammonium ion, NH_4^+.

Salts are formed when acids react with:

- a metal
- a base
- a carbonate
- an alkali

Acid and metal

Balanced equation: $2HCl(aq) + Zn(s) \rightarrow ZnCl_2(aq) + H_2(g)$

Ionic equation: $2H^+(aq) + Zn(s) \rightarrow Zn^{2+}(aq) + H_2(g)$

Acid and carbonate

Balanced equation: $2HCl(aq) + Na_2CO_3(aq) \rightarrow 2NaCl(aq) + H_2O(l) + CO_2(g)$

Ionic equation: $2H^+(aq) + CO_3^{2-}(aq) \rightarrow H_2O(l) + CO_2(g)$

Acid and base

Balanced equation: $2HCl(aq) + MgO(s) \rightarrow MgCl_2(aq) + H_2O(l)$

Ionic equation: $2H^+(aq) + MgO(s) \rightarrow Mg^{2+}(aq) + H_2O(l)$

Acid and alkali

Balanced equation: $HCl(aq) + NaOH(aq) \rightarrow NaCl(aq) + H_2O(l)$

Ionic equation: $H^+(aq) + OH^-(aq) \rightarrow H_2O(l)$

Bases such as ammonia can react with acids to produce a salt. Ammonium sulfate, $(NH_4)_2SO_4$, is used as a fertiliser, and is manufactured by reacting ammonia with sulfuric acid:

Balanced equation: $2NH_3(aq) + H_2SO_4(aq) \rightarrow (NH_4)_2SO_4(aq)$

Ionic equation: $2NH_3(aq) + 2H^+(aq) \rightarrow 2NH_4^+(aq)$

Knowledge check 24

Describe what you would see, if anything, in each of the four reactions of acids:

a acid + metal   **c** acid + base
b acid + carbonate   **d** acid + alkali

Knowledge check 25

Write full equations and ionic equations for each of the following reactions:

a sulfuric acid and solid lithium oxide
b nitric acid and aqueous aluminium hydroxide
c hydrochloric acid and solid barium carbonate

Knowledge check 22

Define an acid, a base and a salt.

Knowledge check 23

a The formula of phosphoric acid is H_3PO_4. What are the formulae of the salts potassium phosphate and barium phosphate?

b The formula of propanoic acid is CH_3CH_2COOH. What are the formulae of the salts sodium propanoate and calcium propanoate?

Exam tip

Many questions ask what you would observe when a certain reaction occurs. Use the state symbols as a guide and remember, you will only see effervescence (bubbles) if a gas is produced.

Acid–base titrations

You will be expected to carry out acid–base titrations, also referred to as volumetric analysis, in the laboratory, and then to undertake structured and unstructured calculations. A titration is a process whereby a precise volume of one solution is added to another solution, until the exact volume required to complete the reaction has been found. With care it is possible to obtain reproducible results accurate to within $0.10\,cm^3$.

Volumetric equipment

Volumetric apparatus includes pipettes, burettes and volumetric flasks.

- A **pipette** is usually selected when a small fixed volume of liquid or solution is required.
- A **burette** is used to provide more variable volumes.
- A **volumetric flask** is used when a larger volume of liquid or solution is required.

Using a pipette

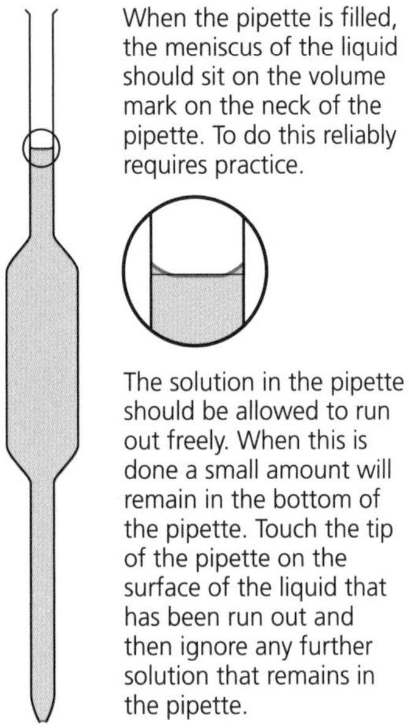

When the pipette is filled, the meniscus of the liquid should sit on the volume mark on the neck of the pipette. To do this reliably requires practice.

The solution in the pipette should be allowed to run out freely. When this is done a small amount will remain in the bottom of the pipette. Touch the tip of the pipette on the surface of the liquid that has been run out and then ignore any further solution that remains in the pipette.

Figure 16 Using a pipette

A pipette is used to deliver a precise volume of solution (Figure 16). The maximum error will vary according to the quality of its manufacture. For routine work a $25.0\,cm^3$ pipette with a maximum error of $0.06\,cm^3$ is often used.

The percentage error is:

$$\frac{0.06}{25.00} \times 100 = 0.24\%$$

This is a much smaller error than would occur if a measuring cylinder had been used. The percentage error using a $25\,cm^3$ measuring cylinder, which has an error of $\pm\,0.5\,cm^3$, is 2.0%, hence the pipette is almost ten times more reliable.

Some pipettes are graduated to allow a volume less than the full capacity to be dispensed. These are useful but usually have a greater maximum error than a standard pipette.

Since the solution sucked into the pipette may be poisonous, a pipette filler must always be used.

Using a burette

When used correctly a volume measured by a burette (Figure 17) has only a small error.

A funnel should be used to fill a burette so that solution is not spilled down the outside. Care must be taken not to overfill it. The tap should, of course, be closed during this procedure. However, before recording a volume, the funnel should be removed and some solution should be run out so that the space below the tap is filled.

The required volume is measured as the difference between the final and initial readings. It is usual to read from the bottom of the meniscus but, because two readings are being used, and one subtracted from the other, the top of the solution could be used for both instead.

Burettes usually have graduations every 0.1 cm³ and should, therefore, have a maximum error of 0.05 cm³. The volume delivered is the difference between two measurements, so the volume obtained has a potential uncertainty of 0.1 cm³. The percentage error will vary with the volume that has been measured.

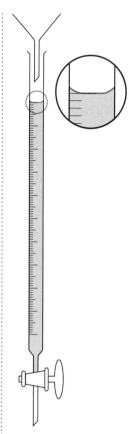

Figure 17 A burette

Knowledge check 26

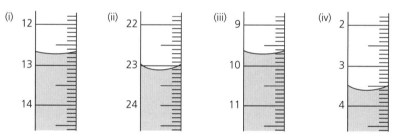

a Record the readings for each of the four burettes.
b Calculate the % error in each of the burette readings. Assume the maximum error of each reading to be 0.05 cm³ and give your answer to 2 decimal places.
c If the burette reading in image (i) is the initial reading and the burette reading in (ii) is the final reading:
  i calculate the titre volume
  ii calculate the % error in the titre volume

Exam tip

There are different grades of burette, so the maximum error in any reading will vary. In an exam the maximum error will be stated if you are required to consider it.

Using a volumetric flask

The mark on the neck of a volumetric flask indicates a specific volume. The maximum error is almost certain to be quoted on the flask and is usually of the order of 0.125%. If a volumetric flask is used to measure 250.0 cm³, the volume obtained has a maximum error of:

$$\frac{0.125}{100} \times 250 = 0.313 \, \text{cm}^3$$

This means that the true volume lies between 249.7 cm³ and 250.3 cm³.

Exam tip

If 250 cm³ were measured using a 50 cm³ burette filled five times, the exercise would not only be more tedious, it would also have a greater error because each time the burette was used there would be an error in the volume measured.

A volume measured in a volumetric flask is very precise. It holds the specified volume when the meniscus of the solution sits on the engraved marking on the neck of the flask.

If it is necessary to dissolve a solid in water to make the solution to place in a volumetric flask, care is needed. It is usual to dissolve the solid first in a container such as a beaker using a volume of distilled water that is less than that required for the solution. A beaker is used so that, if necessary, the contents can be warmed to encourage the solid to dissolve. Once dissolved, the solution must be allowed to cool to room temperature before being transferred to the volumetric flask, using a funnel. Any solution remaining in the beaker is washed into the flask. Distilled water is added carefully to the flask until the meniscus sits on the engraved mark on the neck. With the stopper firmly in place, the flask is inverted a few times to ensure that the solution is completely mixed.

Solutions made up in a volumetric flask are sometimes referred to as **standard solutions** (see Figure 18).

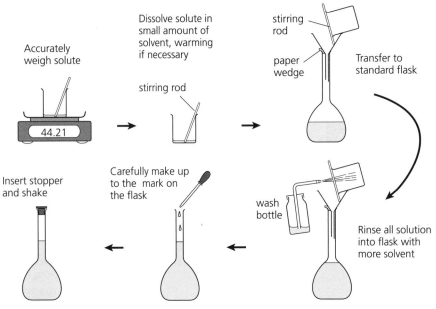

Figure 18

> **Exam tip**
>
> The volume of solution held by a volumetric flask will depend on its temperature, so that hot solutions must not be poured into a volumetric flask.

> A **standard solution** is a solution with a precisely known concentration.

Example

It was found that $18.60\,\text{cm}^3$ HCl(aq) exactly neutralised $25\,\text{cm}^3$ $0.100\,\text{mol}\,\text{dm}^{-3}$ NaOH(aq). Calculate the concentration of the HCl(aq) solution.

Answer

Step 1: Work out how many moles of sodium hydroxide were neutralised. This is possible, as it is the solution for which you know both the concentration used, c, and the volume, V:

$$n = cV = 0.100 \times 0.0250$$

The concentration of the sodium hydroxide is $0.100 \, mol \, dm^3$ of solution. Since $25.0 \, cm^3$ was used, this would contain:

$$0.100 \times \frac{25.0}{1000} = 0.00250 \, mol$$

Step 2: Refer to the balanced equation to see how many moles of hydrochloric acid would be needed to react with this number of moles of sodium hydroxide. The equation for the reaction is:

$$NaOH + HCl \rightarrow NaCl + H_2O$$

The mole ratio is: 1   :   1 :   1 :   1

and therefore 1 mole of NaOH reacts with 1 mole of HCl.

It is known that $0.00250 \, mol$ of NaOH were neutralised, and this must have required $0.00250 \, mol$ of HCl to react completely. Since $18.60 \, cm^3$ of HCl were added from the burette, it must follow that $18.60 \, cm^3$ contained $0.00250 \, mol$ of acid.

Step 3: Convert the information obtained about the hydrochloric acid into its concentration in $mol \, dm^{-3}$.

$$c = \frac{n}{V} = \frac{0.00250}{18.6/1000}$$

$$= \frac{0.00250}{0.0186} = 0.134 \, mol \, dm^3$$

So concentration of HCl was $0.134 \, mol \, dm^{-3}$.

Redox

Redox reactions are reactions in which electrons are transferred from one substance to another. There are numerous everyday examples of redox reactions, including the combustion of fuels and the rusting of iron. Oxidation was originally defined as the gaining of oxygen. For example, when zinc reacts with oxygen to form zinc oxide, the zinc has clearly been oxidised:

$$Zn(s) + \tfrac{1}{2}O_2(g) \rightarrow ZnO(s)$$

A closer inspection of this reaction shows that the zinc atom has been converted to a zinc ion, and in the process has lost two electrons:

$$Zn \rightarrow Zn^{2+} + 2e^-$$

The definition of oxidation therefore has been extended, so that a species is said to be oxidised if it loses electrons. The converse is true for reduction. One way of remembering this is:

OILRIG

Oxidation Is Loss   Reduction Is Gain

> **Knowledge check 27**
>
> If $0.0264 \, mol$ of hydrochloric acid is neutralised by $25.0 \, cm^3$ of sodium carbonate, what amount (in moles) of sodium carbonate is present in the $25.0 \, cm^3$?

Oxidation number

The oxidation number is a convenient way of quickly identifying whether or not a substance has undergone either oxidation or reduction. In order to work out the oxidation number, you must first learn a few simple rules (Table 15).

Rule	Example
All elements in their natural state have an oxidation number zero	H_2; oxidation number of H = 0
The oxidation numbers of the atoms of any molecule add up to zero	H_2O; sum of oxidation numbers = 0
The oxidation numbers of the components of any ion add up to the charge of the ion	SO_4^{2-}; sum of oxidation numbers = –2

Table 15

There are certain elements whose oxidation numbers never change, but some other elements have variable oxidation numbers, and so these have to be deduced.

When calculating the oxidation numbers of elements in either a molecule or an ion, you should apply the following order of priority.

1 Group 1, 2 and 3 elements are always +1, +2 and +3, respectively

2 Fluorine is always –1

3 Hydrogen is usually +1

4 Oxygen is usually –2

5 Chlorine is usually –1

By applying these rules in sequence it is possible to deduce any other oxidation number.

Example 1

Deduce the oxidation number of the chlorine in NaClO.

Answer

The sum of the oxidation numbers in NaClO must add up to zero.

In order of priority, sodium comes first and must be +1; oxygen comes second and is –2. In order for the oxidation numbers to add up to zero, the oxidation number of Cl must be +1. The oxidation number of the chlorine is reflected in the name of NaClO: sodium chlorate(I).

Example 2

Deduce the oxidation number of Cl in the chlorate ion (ClO_3^-).

Answer

The oxidation numbers must add up to the charge on the ion, i.e. –1.

In order of priority, oxygen comes first and is –2, but there are three oxygens, and hence the total is –6. In order for the oxidation numbers to add up to the charge of the ion (–1) the chlorine must be +5. The oxidation number of the chlorine is again shown in the name of the ClO_3^- ion: chlorate(v).

When magnesium reacts with steam, magnesium oxide and hydrogen are formed:

$$Mg(s) + H_2O(g) \rightarrow MgO(s) + H_2(g)$$

It is easy to see that magnesium has been oxidised because it has gained oxygen, and that water has been reduced because it has lost oxygen.

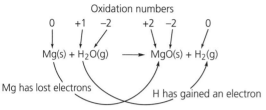

Figure 19

When this reaction is investigated using oxidation numbers (Figure 19), it is clear that an increase in oxidation number is *oxidation*, while a decrease in oxidation number is *reduction*.

Electron transfer can be shown by using half-equations, such that:

$$Mg \rightarrow Mg^{2+} + 2e^- \qquad (loss \text{ of electrons is oxidation})$$

$$2H^+ + 2e^- \rightarrow H_2 \qquad (gain \text{ of electrons is reduction})$$

A number of elements can have more than one oxidation number. Iron reacts with chlorine and it is possible to form two different chlorides, $FeCl_2$ and $FeCl_3$, so that Fe can have either an oxidation number of 2 or an oxidation number of 3. Since there is a possibility of confusion, the name given to the compounds must make the oxidation number absolutely clear. Iron(II) chloride ($FeCl_2$) is so named to indicate that it is a compound containing iron with a valency of 2, while iron(III) chloride makes it clear that we are referring to $FeCl_3$.

Redox reactions

Metals generally react by losing electrons (oxidation is loss), causing the oxidation number to increase (Figure 20).

Oxidation numbers $\qquad$ 0 $\qquad\qquad$ +2

$$Mg \longrightarrow Mg^{2+} + 2e^-$$

Figure 20

Non-metals generally react by gaining electrons (reduction is gain), causing the oxidation number to decrease (Figure 21).

Oxidation numbers $\qquad$ 0 $\qquad\qquad$ −1

$$F_2 + 2e^- \longrightarrow 2F^-$$

Figure 21

> **Exam tip**
>
> If asked about a redox reaction it is good practice to work out the oxidation number of each element in the equation. Write the oxidation numbers above each element and check to see if any oxidation numbers have increased (oxidation) or decreased (reduction). Remember you cannot have one without the other.

> **Knowledge check 28**
>
> Deduce the oxidation number of the underlined element in each of the following:
>
> $Na_2\underline{S}O_3$ $\underline{Cl}O_4^-$ $\underline{Cr}O_4^{2-}$
>
> Suggest a name for each substance.

When a metal reacts with either HCl(aq) or H_2SO_4(aq), the metal is oxidised (Figure 22).

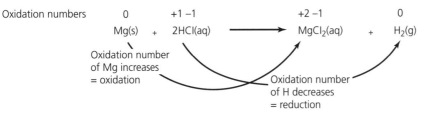

Figure 22

You should be able to use oxidation numbers to determine what has been oxidised or reduced in unfamiliar reactions, such as:

$$4HCl + MnO_2 \rightarrow MnCl_2 + Cl_2 + 2H_2O$$

Work out the oxidation number of each element on both sides of the equation. Identify the element whose oxidation number has *increased* (oxidation) and the element whose oxidation number has *decreased* (reduction) — see Figure 23.

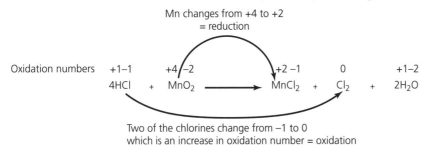

Figure 23

Summary

Having revised Module 2 Section 1: **Atoms and reactions**, you should now have an understanding of:

- atomic structure, isotopes, relative masses and mass spectra
- deducing formulae and writing equations
- mole calculations using the equations

$$n = \frac{m}{M}, \; n = \frac{V \text{ (in dm}^3\text{)}}{24} \text{ and } n = cV$$

- calculating the formulae of hydrated salts
- acids, bases and the formation of salts
- titrations
- oxidation and reduction in terms of oxidation numbers

Electrons, bonding and structure

Electron structure

Ionisation energy provides evidence for the existence of electron shells and sub-shells.

You should be able to define the **first ionisation energy**.

The first ionisation energy can be represented by the equation:

$$X(g) \rightarrow X^+(g) + e^-$$

It is important to include the state symbols.

For elements that have more than one electron, it is possible to remove electrons one by one from the atom. The **second ionisation energy** starts with a 1+ ion and results in the formation of a 2+ ion.

The second ionisation energy can be represented by the equation:

$$X^+(g) \rightarrow X^{2+}(g) + e^-$$

The third ionisation energy will form a 3+ ion and start with a 2+ ion. It follows that the eighth ionisation energy is:

$$X^{7+}(g) \rightarrow X^{8+}(g) + e^-$$

Energy levels, shells and sub-shells

There are three factors that influence ionisation energy:

- the distance of the outermost electron from the nucleus (atomic radius)
- electron shielding (the number of inner electron shells)
- nuclear charge (the number of protons in the nucleus)

As we go across a period, the atomic radii decrease and the main shell shielding remains the same. This should make it more difficult to remove an electron. The nuclear charge also increases across a period, making it more difficult to remove an electron. Therefore, ionisation energy increases across a period.

As we go down a group, atomic radii and shielding both increase, and this should make it easier to remove an electron, but as we go down a group the nuclear charge also increases, which should make it harder to remove an electron. Increases in atomic radii and shielding outweigh nuclear charge, so that ionisation energy decreases down a group.

It is possible to remove each electron, one by one, from an atom, and to measure the size of each successive ionisation energy. When the successive energies are plotted, the graph provides evidence for the existence of shells. Chlorine has 17 electrons, which we know to be arranged as 2,8,7 in three electron shells. The evidence for this is shown in a plot of successive ionisation energies (Figure 24).

The **first ionisation energy** is the energy required to remove one electron from each atom in 1 mole of gaseous atoms.

The **second ionisation energy** is the energy required to remove one electron from each ion with 1+ charge in 1 mole of gaseous ions.

Knowledge check 30

Write an equation to show the fourth ionisation energy of Ca(s).

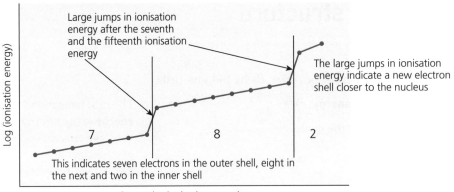

Figure 24 Successive ionisation energies of $_{17}Cl$

Plotting successive ionisation energies confirms what you learnt at GCSE. It shows clearly that the first shell contains a maximum of two electrons and the second shell contains a maximum of eight electrons. There is further experimental evidence to suggest that each shell is made of smaller sub-shells.

By studying the first ionisation energy of the first 20 elements, evidence for the existence of sub-shells was obtained (Figure 25).

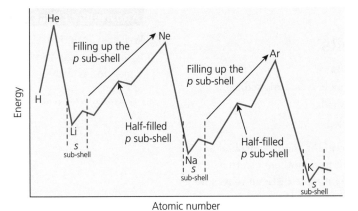

Figure 25

The trends in ionisation energy suggested that each group 1 element would have a low first ionisation energy and that this ionisation energy would increase across the period to reach a maximum at group 8 (the noble gases). There is a gradual increase in ionisation energies across a period, but several small peaks and troughs exist. These peaks and troughs are repeated in each period. They provide evidence for the existence of sub-shells (s, p and d). There is a periodic variation throughout all of the elements and there is evidence for several shells and sub-shells.

Shell	Sub-shells				Number of electrons in each sub-shell			
First	1s				2			
Second	2s	2p			2	6		
Third	3s	3p	3d		2	6	10	
Fourth	4s	4p	4d	4f	2	6	10	14

Table 16

We now know that the sub-shells are made up of orbitals.

Atomic orbitals

The concept of an **orbital** is difficult. If you are asked in an exam to define an orbital, the simplest explanation is 'a region around the nucleus where there is a high probability of finding an electron. Each orbital can hold up to a maximum of two electrons, each spinning in opposite directions.' You should be able to describe, with the aid of a diagram, the shape of *s*- and *p*-orbitals (Figure 26).

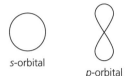

s-orbital   *p*-orbital

Figure 26

There are three *p*-orbitals, one along each of the *x*-, *y*- and *z*-axes (Figure 27).

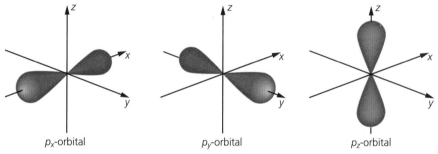

p_x-orbital p_y-orbital p_z-orbital

Figure 27

The sequence in which electrons fill the orbitals is shown in the energy diagram in Figure 28.

Remember that electrons are represented by arrows:

- within an orbital, the electrons must have opposite spins
- the lowest energy level is occupied first
- orbitals at the same energy level are occupied singly before pairing of electrons

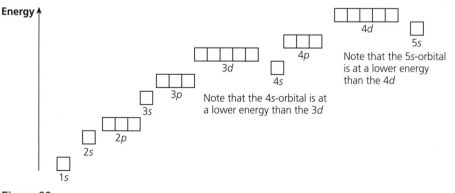

Figure 28

An **orbital** is a region around the nucleus that can hold up to two electrons, with opposite spins.

Electron configuration

You should be able to write the full electronic configuration of the first 36 elements in the following manner:

- $_{17}Cl$ is $1s^2\, 2s^2\, 2p^6\, 3s^2\, 3p^5$ or $[Ne]\, 3s^2\, 3p^5$
- $_{26}Fe$ is $1s^2\, 2s^2\, 2p^6\, 3s^2\, 3p^6\, 3d^6\, 4s^2$ or $[Ar]\, 3d^6\, 4s^2$

You should also be able to write the full electronic configuration of the ions of the first 36 elements in the following manner:

- $_{17}Cl^-$ is $1s^2\, 2s^2\, 2p^6\, 3s^2\, 3p^6$ or $[Ne]\, 3s^2\, 3p^6$
- $_{26}Fe^{2+}$ is $1s^2\, 2s^2\, 2p^6\, 3s^2\, 3p^6\, 3d^6$ or $[Ar]\, 3d^6$

It is a common mistake to write $_{26}Fe$ as $1s^2\, 2s^2\, 2p^6\, 3s^2\, 3p^6\, 4s^2\, 3d^6$ and then remove the $3d$ electrons first, such that $_{26}Fe^{2+}$ is written as $1s^2\, 2s^2\, 2p^6\, 3s^2\, 3p^6\, 4s^2\, 3d^4$. This is *not* the case and it will cost you marks in the exam. It is worth remembering that the $3d$ electrons are part of an inner shell and the outer shell ($4s$) electrons will always be lost first.

The electronic configuration can also give a clue as to relative stability. Comparing the electronic configuration of Fe^{2+} and Fe^{3+}:

- $_{26}Fe^{2+}$ is $1s^2\, 2s^2\, 2p^6\, 3s^2\, 3p^6\, 3d^6\, 4s^2$ or $[Ar]\, 3d^6$
- $_{26}Fe^{3+}$ is $1s^2\, 2s^2\, 2p^6\, 3s^2\, 3p^6\, 3d^5\, 4s^2$ or $[Ar]\, 3d^5$

This shows that Fe^{3+} has a half-filled d shell ($3d^5$), while one of the d-orbitals in Fe^{2+} is occupied by two electrons and hence is not as stable as Fe^{3+}.

The full electronic configuration allows you to classify elements into s-, p- and d-block elements and to identify their positions in the periodic table. See, for example, Figure 29.

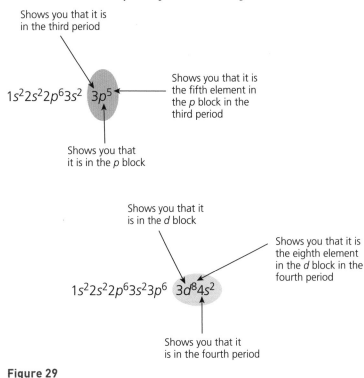

Shows you that it is in the third period

$1s^2 2s^2 2p^6 3s^2$ $3p^5$

Shows you that it is the fifth element in the p block in the third period

Shows you that it is in the p block

Shows you that it is in the d block

Shows you that it is the eighth element in the d block in the fourth period

$1s^2 2s^2 2p^6 3s^2 3p^6$ $3d^8 4s^2$

Shows you that it is in the fourth period

Figure 29

Exam tip

When filling up the orbitals the $4s$ fills before the $3d$ *but* when electrons are lost the $4s$ ones leave before the $3d$ ones. It is *not* last in first out. The $4s$ are lost before the $3d$ because they are in the outer 4th shell while the $3d$ electrons are in the inner 3rd shell.

Knowledge check 32

Write the full electronic configurations of $_{24}Cr$ and $_{23}V^{2+}$.

Bonding and structure

Types of bond

There are three main types of bond: **ionic**, **covalent** and **metallic**. Table 17 summarises each type.

	Ionic	Covalent	Metallic
Definition	An ionic bond is the electrostatic attraction between oppositely charged ions	A covalent bond is the strong electrostatic attraction between a shared pair of electrons and the nuclei of the bonded atoms	A metallic bond is the electrostatic attraction between positive metal ions in the lattice and delocalised electrons
Formation	Formed by electron transfer from metal atom (X) to non-metal atom (Y) to produce oppositely charged ions X^+ and Y^-	Formed when electrons are shared rather than transferred	The positive ions occupy fixed positions in a lattice and the delocalised electrons can move freely throughout the lattice
Direction	An ionic bond is directional, acting between adjacent ions	A covalent bond is directional, acting solely between the two atoms involved in the bond	A metallic bond is non-directional, because the delocalised electrons can move anywhere in the lattice
Examples	NaCl, MgO	H_2, CH_4	Cu, Na
Melting and boiling points	High melting point and boiling point due to strong electrostatic forces between ions throughout the solid lattice	Low melting point and boiling point — the simple covalent molecules are held together by weak forces between the molecules and hence little energy is needed to break the weak intermolecular forces	High melting point and boiling point due to strong metallic bonds between positive ions and the delocalised electrons throughout the lattice
Conductivity	Non-conductor of electricity in solid state, but conducts when melted or dissolved in water because the ionic lattice breaks down and ions are free to move as mobile charge carriers	Non-conductors of electricity — no free or mobile charged particles	Good thermal and electrical conductors due to mobile, delocalised electrons which conduct heat and electricity, even in the solid state
Solubility	The ionic lattice usually dissolves in polar solvents (for example, water) Polar water molecules attract ions in lattice and surround each ion (hydration)	Simple molecular structures soluble in non-polar solvents (for example, hexane), but usually insoluble in water	Insoluble in polar and non-polar solvents. Some metals (groups 1 and 2) react with water

Table 17

In addition to covalent bonding, **dative covalent bonds** or **coordinate bonds** exist. These also are the result of two shared electrons, but in this case only one of the atoms supplies *both* shared electrons.

A **dative covalent bond** is defined as the strong electrostatic attraction between a shared pair of electrons and the nuclei of the bonded atoms, where one of the bonded atoms supplies both shared electrons.

Dot-and-cross diagrams

Dot-and-cross diagrams are a simple visual way to illustrate both ionic and covalent bonding (Figure 30).

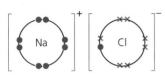

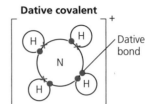

Ionic	Covalent	Dative covalent
Must use a 'dot' to show transfer of an electron	Each bond is made up of a 'dot' and a 'cross'	The dative bond is made up of two 'dots'

Figure 30

Knowledge check 33

a State the type of bonding in:
  i CuO   ii P_2O_5 iii Mo   iv $SOCl_2$ v $SrCl_2$
b Draw dot-and-cross diagrams of:
  i CaO   ii SiF_4

The shapes of molecules and ions

The **electron-pair repulsion theory** states that the shape of a covalent molecule is determined by the number and nature of electron pairs around the central atom. The electron-pair repulsion theory depends on the behaviour of these electron pairs, such that:

- Electron pairs repel one another so that they are as far apart as possible.
- Lone pairs of electrons are more 'repelling' than bonded pairs of electrons.

You should be able to draw a dot-and-cross diagram of a molecule and use it to determine the number and type of electron pairs around the central atom. You can then use Table 18 to predict the shape and the bond angle (Figure 31).

Number of bonded pairs of electrons	Number of lone pairs of electrons	Shape	Approximate bond angle
2	0	Linear	180°
3	0	Trigonal planar	120°
4	0	Tetrahedral	109.5°
6	0	Octahedral	90°
5	0	Trigonal bipyramidal	90° and 120°
3	1	Pyramidal	107°
2	2	Angular	104°

Table 18

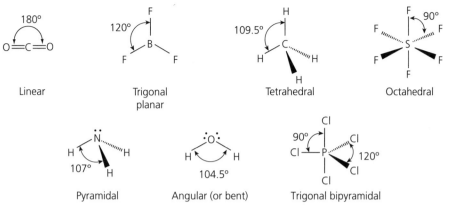

Figure 31

Knowledge check 34

Draw dot-and-cross diagrams of PH_3, NH_2^- and $COCl_2$. Use your diagrams to decide the shape and the bond angles in each.

Electronegativity and bond polarity

Ionic and covalent bonding are extremes — there is a range of intermediate degrees of bonding between. Many bonds are described as:

- essentially covalent with some ionic character, due to differences in **electronegativity** — bond polarity
- essentially ionic with some covalent character, due to differences in **charge density** — polarisation

If a covalent bond is formed between two different elements it is highly likely that each of the elements will attract the covalent bonded pair of electrons unequally.

Electronegativity

Electronegativity increases across a period but decreases down a group, so that, for example, fluorine is the most electronegative element (Figure 32).

Electronegativity is the strength of the attraction of an atom, in a molecule, for the pair of electrons in a covalent bond.

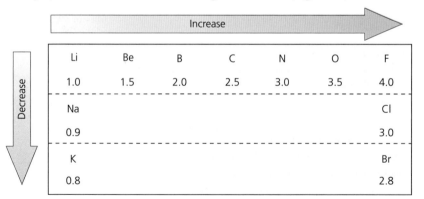

Figure 32

In molecules like hydrogen chloride (HCl), the two electrons in the covalent bond are shared unequally. Chlorine has a higher electronegativity than hydrogen, so that the two shared electrons are pulled towards the chlorine atom, resulting in the formation of a permanent dipole. The bond in HCl consists of two shared electrons (essentially covalent), but also contains $\delta+$ and $\delta-$ charges (hence the ionic character).

- The greater the *difference* between electronegativities, the greater is the *ionic* character of the bond.
- The greater the *similarity* in electronegativities, the greater is the *covalent* character of the bond.

Polarity

If a compound consists of two or more different non-metals bonded together, the bonds are most likely to be essentially covalent with some ionic character, such that the bonds have dipoles. However, if the molecule is symmetrical the dipoles cancel, resulting in a non-polar molecule. If you refer to Figure 31 all of the shapes, except 'pyramidal' and 'angular', are symmetrical and any dipoles in their bonds cancel, resulting in the formation of a non-polar molecule. The only polar molecules in Figure 31 are ammonia and water. Linear molecules, such as HCl, which are formed between two atoms with different electronegativities, are also polar (Figure 33).

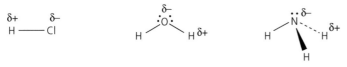

Figure 33

Exam tip

If the central atom in a covalent molecule only has bonded pairs of electrons the overall molecule will be symmetrical and the dipoles will cancel such that the molecule will be non-polar.

Intermolecular forces

The three main categories of bonds (ionic, covalent and metallic) described are all strong: 200–$600\,\text{kJ}\,\text{mol}^{-1}$ of energy are needed to break them. There is a group of other 'bonds' known collectively as **intermolecular forces**, which are weak bonds that need only 2–$40\,\text{kJ}\,\text{mol}^{-1}$ of energy to break them. The intermolecular forces are:

- permanent dipole–dipole interactions
- induced dipole–dipole interactions
- hydrogen bonds

Permanent dipole–dipole interactions

These interactions are usually found between polar molecules that are essentially covalent but have some ionic character. They are weak intermolecular forces between the permanent dipoles of adjacent molecules (Figure 34).

Induced dipole–dipole interactions

Induced dipole–dipole interactions are the weakest of the intermolecular forces and act between all particles, polar or non-polar. They are caused by the movement of electrons. This movement results in an oscillating or instantaneous dipole, which in turn induces a dipole on neighbouring molecules. The attraction between these induced dipoles produces weak intermolecular interactions. The strength of the intermolecular interactions depends on the number of electrons in the molecule. The greater the number of electrons in an atom or molecule, the greater are the induced dipole–dipole interactions.

Intermolecular forces are weak forces of attraction *between* covalent molecules.

Weak force of attraction between the $Cl^{\delta-}$ in one HCl and the $H^{\delta+}$ in the next HCl

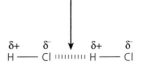

Figure 34

If you are asked to explain or define an induced dipole–dipole interaction, there are three key features that you must include:

- the movement of electrons generates an instantaneous dipole
- the instantaneous dipole induces another dipole in neighbouring atoms or molecules
- the attraction between the temporary induced dipoles results in the induced dipole–dipole interactions

Hydrogen bonds

Hydrogen bonds exist between molecules that contain hydrogen that is bonded to nitrogen, oxygen or fluorine. They are comparatively strong permanent dipole–dipole interactions, involving a lone pair of electrons on either the nitrogen, oxygen or fluorine atom. Hydrogen bonds exist between molecules of ammonia, water (Figure 35) and hydrogen fluoride. They are also found in alcohols.

Special properties of water arise from hydrogen bonding:

- The solid (ice) is less dense than the liquid (water) because the hydrogen bonds in ice hold the H_2O molecules further apart, creating a more open lattice structure.
- Water has a relatively high melting point and boiling point, due to the additional energy required to break the hydrogen bonds.

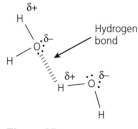

Figure 35

Exam tip

Make sure that you can correctly draw the formation of H-bonds in molecules other than water. You may be asked to illustrate the formation of H-bonds in molecules such as ammonia, alcohols and possibly amines such as CH_3NH_2. The H-bond in ammonia is illustrated in Figure 36.

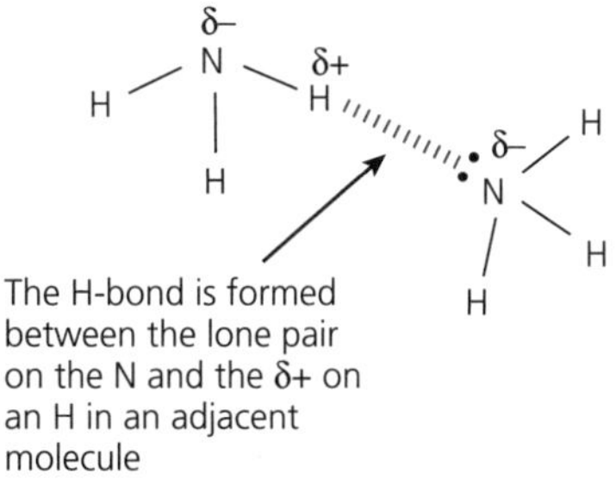

The H-bond is formed between the lone pair on the N and the δ+ on an H in an adjacent molecule

Figure 36

Knowledge check 35

Alcohols such as methanol, CH_3–O–H, also form H-bonds. Draw two molecules of methanol and show the formation of an H-bond. Explain why methanol has a higher than expected boiling point.

Structure, bonding and physical properties

Giant ionic structures are held together by strong electrostatic attractions between the ions throughout the lattice. Properties of giant ionic structures include:

- high melting points and boiling points
- they are good conductors when molten or in an aqueous solution but not when solid, because only when they are molten or dissolved in water do they have mobile charged particles (ions)
- they are soluble in polar solvents such as water

Simple covalent structures are molecular and the molecules are held together by weak intermolecular forces. However, some covalent molecules, such as water and iodine, also form **simple molecular lattices**. The molecules are held in position in the lattice by intermolecular forces that are comparatively easy to break. Properties include:

- low melting points and boiling points
- they are poor conductors because they do not have any mobile charged particles (electrons or ions)
- they are soluble in non-polar solvents such as hexane

Summary

Having revised Module 2 Section 2: **Electrons, bonding and structure** you should now have an understanding of:

- ionisation energies
- electron configuration using $1s^2, 2s^2$... notation

- ionic, covalent and metallic bonding
- shapes of covalent molecules and ions
- electronegativity and bond polarity
- intermolecular forces including hydrogen bonding and van der Waals forces

Questions & Answers

Approaching the exam

There are four modules in the AS (H032) specification and a further two modules in the A-level (H432) specification. The table summarises the structure and content of the exams for AS and for A-level.

All questions used in this book are relevant to both AS and A-level examinations unless otherwise stated.

	Component exam	Modules						Total marks/ time	Type of questions
		1	2	3	4	5	6		
AS	Breadth in chemistry	✓	✓	✓	✓			70 marks 1 h 30 min	Multiple choice (20 marks) Short answer questions (50 marks)
	Depth in chemistry	✓	✓	✓	✓			70 marks 1 h 30 min	Short answer and extended response questions (70 marks)
A-level	Periodic table, elements and physical chemistry	✓	✓	✓		✓		100 marks 2 h 15 min	Multiple choice (15 marks) Short answer and extended response questions (85 marks)
	Synthesis and analytical techniques	✓	✓		✓		✓	100 marks 2 h 15 min	Multiple choice (15 marks) Short answer and extended response questions (85 marks)
	Unified chemistry	✓	✓	✓	✓	✓	✓	70 marks 1 h 30 min	Short answer and extended response questions (70 marks)

Question types

Multiple-choice questions will be used in AS paper 1 and in A-level papers 1 and 2. Multiple-choice questions need to be read carefully and there is often a misconception that these questions have to be done 'in your head'. This is not the case. Many multiple-choice questions require you to think and to work things out on paper, and space will be provided on the question paper. For each of the questions there are four suggested answers: A, B, C or D. You select your response by putting a cross in the box by the letter of your choice. Multiple-choice questions are machine-marked and it is essential that you follow the instructions given on the exam paper. Typical multiple-choice questions are illustrated in this book by questions 1–10. The multiple-choice questions in this book will not be 'machine-marked' and therefore do not follow the exam format exactly.

Short answer questions will appear in all examinations at both AS and at A-level. Read the questions carefully, be aware of the marks awarded for each section — for example if there are 2 marks for a sub-section in a question you will be expected to make two points. The space provided for the answer has been designed to be more than sufficient for a complete response. Do not write in the margins. If you do require additional space, ask for extra paper and explain on the question paper that you have used extra paper for this response. Typical short answer questions are illustrated in this book by questions 11–19.

Extended response questions will appear on paper 2 at AS and on all three papers at A-level. Students often think of this type of question as 'essay' questions — but you don't have to use an essay style when answering these questions. Chemists communicate in a variety of ways, such as formulae, equations and mechanisms, all of which are difficult to put into an essay style. It is perfectly acceptable to use bullet points or to tabulate your response. Extended response questions require that you plan your answer carefully. Make sure that you use the marks allocated as a guidance — if there are 8 marks you must make eight separate points. Check the mark schemes from previous papers and you will see that each mark is allocated to a specific point. Typical extended response questions are illustrated in this book by questions 20–22.

Terms used in exam questions

You will be asked precise questions in the examinations, so you can save a lot of valuable time (as well as ensuring that you score as many marks as possible) by knowing what is expected. Terms most commonly used are explained below.

Define: this is intended literally. Only a formal statement or equivalent paraphrase is required.

Explain: this normally implies that a definition should be given, together with some relevant comment on the significance or context of the term(s) concerned, especially where two or more terms are included in the question. The amount of supplementary comment intended should be determined by the mark allocation.

State: this implies a concise answer with little or no supporting argument.

Describe: this requires you to state in words (using diagrams where appropriate) the main points of the topic. It is often used with reference either to particular phenomena or to particular experiments. In the former instance, the term usually implies that the answer should include reference to (visual) observations associated with the phenomena. The amount of description intended should be interpreted according to the indicated mark value.

Deduce or predict: this implies that you are not expected to produce the required answer by recall but by making a logical connection between other pieces of information. Such information may be wholly given in the question or may depend on answers given in an earlier part of the question. 'Predict' also implies a concise answer, with no supporting statement required.

Outline: this implies brevity, that is, restricting the answer to essential detail only.

Suggest: this is used in two main contexts. It implies either that there is no unique answer or that you are expected to apply your general knowledge to a 'novel' situation that may not be formally in the specification.

Calculate: this is used when a numerical answer is required. In general, working should be shown.

Sketch: when applied to diagrams this implies that a simple, freehand drawing is acceptable. Nevertheless, care should be taken over proportions, and important details should be labelled clearly.

About this section

This section contains questions similar in style to those you can expect to find in your AS and A-level examinations. Each question in this section identifies the specification topic, the total marks and a suggested time that should be spent writing out the answer.

The limited number of questions means that it is impossible to cover all the topics and question styles, but they should give you a flavour of what to expect. The responses that are shown are real students' answers to the questions. Student A is an A/B-grade student and student B is a B/C-grade student.

Apart from question 22, which is for A-level students only, the questions are relevant to both AS and A-level students.

There are several ways of using this section. You could:

■ 'hide' the answers to each question and try the question yourself. It needn't be a memory test — use your notes to see if you can make all the necessary points
■ check your answers against the students' responses and make an estimate of the likely standard of your response to each question
■ take on the role of the examiner and mark each student's response, then check whether you agree with the marks awarded
■ check your answers against the comments to see if you can appreciate where you might have lost marks

Exam advice

Comments on the questions are preceded by the icon **ⓔ**. They offer tips on what you need to do in order to gain full marks. All student responses are followed by comments highlighting where credit is due, indicated by the icon **ⓔ**. In the weaker answers, they also point out areas for improvement, specific problems and common errors such as lack of clarity, irrelevance, misinterpretation of the question and mistaken meanings of terms.

Multiple-choice questions

Questions in this section are relevant to AS Component 1 and to A-level Components 1 and 2.

Answer the questions that follow and record your answers. Check your answers when you have completed all 10 questions.

Question 1

Which one of the following has more neutrons than electrons **and** more electrons than protons?

A $^{19}_{9}F^{-}$

B $^{80}_{35}Br^{-}$

C $^{9}_{4}Be$

D $^{9}_{4}Be^{2+}$

Question 2

Successive ionisation enthalpies of a period 3 element, X, in $kJ\,mol^{-1}$ are:

789, 1577, 3232, 4356, 16091, 19785, 23787, 29253

What is X?

A fluorine

B carbon

C chlorine

D silicon

Question 3

Which of the following is the correct formula of thorium(IV) sulfate? (The symbol for thorium is Th.)

A $Th_2(SO_4)_4$

B $Th(SO_4)_2$

C $Th(SO_4)_4$

D Th_4SO_4

Question 4

Which one of the following contains a number of particles equal to the Avogadro constant?

A molecules in 14.0 g of nitrogen gas

B atoms in 14.0 g of silicon

C potassium ions in 1 mol of potassium sulfate

D carbonate ions in 1 mol of barium carbonate

Question 5

What volume of $0.50\,mol\,dm^{-3}$ nitric acid will react with $20.0\,cm^3$ of $0.40\,mol\,dm^{-3}$ lithium carbonate?

A $10.0\,cm^3$

B $16.0\,cm^3$

C $32.0\,cm^3$

D $40.0\,cm^3$

Question 6

What is the shape of a molecule of phosphorus trichloride, PCl_3?

A trigonal planar   **B** tetrahedral

C pyramidal   **D** angular

Question 7

Which type of bonding is present in the ammonium ion, NH_4^+?

A ionic

B covalent

C metallic

D essentially covalent but with some ionic character (polar)

Use Table 1 to answer questions 8, 9 and 10.

A	B	C	D
1, 2 and 3 correct	1, 2 correct	2, 3 correct	3 only correct

Table 1

Question 8

The correct ionic equation for the reaction of solid strontium carbonate with hydrochloric acid is:

$$SrCO_3(s) + 2H^+(aq) \rightarrow Sr^{2+}(aq) + CO_2(g) + H_2O(l)$$

One of more of the statements 1–3 may be correct. Select A, B, C or D from Table 1.

1 The strontium carbonate is written as its full formula, $SrCO_3(s)$, as its ions are not free to move.

2 The carbon dioxide is written as its full formula as it does not contain ions.

3 The chloride ion from the hydrochloric acid is excluded from the equation because it is not present when the reaction has finished.

Question 9

Which of the following statements is true for all the elements in the third period of the periodic table? Select answer A, B, C or D from Table 1.

1 They all have electrons in at least six different orbitals.

2 They all have some p-orbitals in their electron structures.

3 Only two of the elements have electron structures with all their orbitals containing pairs of electrons.

Question 10

Which of the following underlined elements have oxidation state 6? Select answer A, B, C or D from Table 1.

1 Na$\underline{I}O_4$

2 $\underline{S}O_3^{2-}$

3 Na$_2\underline{Cr}_2O_7$

Answers to multiple-choice questions

The student answers in the table *are all incorrect* and are used to illustrate the most common incorrect responses.

	1	2	3	4	5	6	7	8	9	10
Student answer	A	B	A	C	B	A	A	A	B	A
Correct answers	B	D	B	D	C	C	B	B	A	D

ⓔ It should now be apparent that you cannot answer all multiple-choice questions in your head — many require working out on paper.

Q1 There are two requirements and you need to check both. It is difficult to keep all eight values in your head and you must make notes in the space provided on the exam paper.

Q2 To work out the group, look for the largest jump in ionisation energy, which in this case is between the 4th and 5th IE, showing that the element is in group 4. You also must remember that the element is in period 3.

Q3 All formulae are given in their simplest (empirical) form. By using the information given and your knowledge of the sulfate ion the formula is Th$(SO_4)_2$.

Q4 The words 'molecules, atoms and ions' and their related formulae are important and you must read each statement carefully and apply it to the formula of the substance.

Q5 This cannot be done without writing an equation and then carrying out the calculation.

Q6 Option A has been deliberately put first to tempt you to simply look at the formula PCl_3 and opt for trigonal planar. Once again it is essential that you take your time and draw a dot-and-cross diagram of PCl_3 before deciding on the answer.

Q7 Again, option A has been put first to tempt you to opt for ionic bonding in the ammonium ion, but the bonding *within* the ammonium ion is covalent.

Q8–10 You must consider each statement separately and decide if the statement is true or false.

Short response questions

Questions in this section are relevant to AS Components 1 and 2 and to A-level Components 1, 2 and 3.

Question 11 Isotopes; electron configuration; ionisation energy

Time allocation: 12–15 minutes

A sample of potassium contains two isotopes: ^{39}K and ^{41}K.

(a) (i)   State what is meant by the term 'isotope'. (1 mark)

   (ii) What is the difference between the two isotopes of potassium? (1 mark)

   (iii) The relative atomic mass of the potassium is 39.1. Define the term 'relative atomic mass'. (2 marks)

   (iv) Calculate the percentage of each isotope of potassium, given that the relative atomic mass is 39.1. (3 marks)

(b) (i)   Write the full electronic configuration of a potassium atom. (1 mark)

   (ii) Define the first ionisation energy of potassium. (3 marks)

   (iii) Write an equation, including state symbols, for the second ionisation of potassium. (2 marks)

   (iv) Explain why the first ionisation energies of ^{39}K and ^{41}K are the same. (1 mark)

(c) The first and second ionisation energies of potassium are 419 and 3051 $kJ\,mol^{-1}$ respectively. Explain why there is a large difference between the first and the second ionisation energies of potassium. (3 marks)

Total: 17 marks

ⓔ The command words 'state' and 'define' used in parts (a)(i) and (b)(ii) indicate a brief answer is required with no supporting argument. The command word 'explain', for parts (b)(iv) and (c), requires a more detailed answer with commentary.

Student A

(a) (i)   Isotopes are atoms of the same element that have the same number of protons but a different number of neutrons.

Student B

(a) (i)   Same number of protons, different number of neutrons.

ⓔ Each student gains the mark. Student A has written a sentence, paying due regard to spelling, punctuation and grammar. However, in such questions, marks are not awarded for quality of written communication (QWC), so you can be brief and just stick to the facts.

Student A

(a) (ii) ^{41}K has two more neutrons than ^{39}K.

Student B

(a) (ii) ^{41}K is heavier than ^{39}K.

ⓔ Student A scores the mark. Student B is unlikely to score because he/she has missed out simple quantitative detail. Always try to be as precise as possible. If numbers are given in a question, you usually have to use them in your answer.

Student A

(a) (iii) The relative atomic mass is the weighted mean mass of an atom of the element compared with that of the mass of an atom of carbon-12, which is taken as exactly 12.

Student B

(a) (iii) It is the average mass based on the carbon-12 scale.

ⓔ Student A has learnt the definition of relative atomic mass and gains both marks. The definition given by Student B does not take into account the amount of each isotope. If this definition is applied to this sample of potassium, then the average mass of the two isotopes (one with mass 39 and the other 41) is 40, not 39.1. Student B scores 1 mark.

Student A

(a) (iv) $(39x) + 41(100 - x) = 100 \times 39.1 = 3910$

$39x + 4100 - 41x = 3910$

Therefore:

$-2x = 3910 - 4100 = -190$

so $x = 95$

The mixture contains 95% ^{39}K and 5% ^{41}K

Student B

(a) (iv) The difference between 39 and 41 is 2, the average mass is 39.1 so the mixture has to contain $\dfrac{19}{20}$ of ^{39}K and $\dfrac{1}{20}$ ^{41}K.

e Both students have answered this difficult question well. Student A has been systematic and scores all 3 marks. Student B has approached the question in a different, but equally correct, way, and has correctly worked out the fraction of each isotope. Unfortunately, student B didn't convert the fractions into percentages and so scores just 2 marks.

> **Student A**
>
> **(b) (i)** $1s^2 \, 2s^2 \, 2p^6 \, 3s^2 \, 3p^6 \, 4s^1$

> **Student B**
>
> **(b) (i)** 2, 8, 8, 1

e Student A scores the mark, but student B does not. The full electronic configuration requires subshells as well as principal shells.

> **Student A**
>
> **(b) (ii)** The first ionisation energy is the energy required to remove one electron from one mole of atoms in the gaseous state at STP.

> **Student B**
>
> **(b) (ii)** It is the loss of an electron from the atom in the gaseous state.

e Student A gains 2 of the 3 marks available; student B scores only 1 mark. Student A implies that only a single electron would be removed from 1 mol (6.02×10^{23}) of atoms. Writing an equation can help to put the key points into words. Student B understands the concept but has not bothered to learn the detail. This can be costly.

> **Student A**
>
> **(b) (iii)** $K^+(g) \rightarrow K^{2+} + e^-$

> **Student B**
>
> **(b) (iii)** $K(g) \rightarrow K^{2+}(g) + 2e^-$

e Each student scores only 1 mark. Student A has been careless and has lost a mark by forgetting to write the state symbol (g) after K^{2+}. Student B earns 1 mark for the state symbols. However, the equation is incorrect because it includes both the first and second ionisations.

> **Student A**
>
> **(b) (iv)** They have the same electron arrangement and same number of protons.

Student B

(b) (iv) The only difference is the number of neutrons and they are neutral.

ⓔ This is a difficult question. Each student's response is true but neither fully addresses the question, in that ionisation energy depends on overcoming the attraction between the electron and the nucleus. This attraction is the same for both isotopes of potassium. Neither student scores the mark.

Student A

(c) The first electron is easy to remove because it is further from the nucleus and is shielded by an extra inner shell. The second ionisation energy is high because the potassium now has a stable noble gas configuration.

Student B

(c) The first ionisation is easy because the electron is further from the nucleus and has more shielding. The second electron is harder to remove because the nucleus is bigger.

ⓔ Examiners are looking for three key points: distance from the nucleus, shielding and how these affect the attraction between the nucleus and the electrons. The final key point is often either not included or is misinterpreted by stating that 'the nucleus gets bigger' — the nucleus remains the same size and the number of electrons changes. The best way to answer this question is to compare the potassium atom, K, with the potassium ion, K^+. The K^+ ion is smaller than the potassium atom, hence the electrons are held more tightly. The K^+ ion has fewer inner shells/less shielding than the K atom, hence the electrons are held more tightly. Both students score 2 marks.

ⓔ **Student A scores 13 out of 17 marks and student B scores 8 marks. If student A maintained this standard throughout an examination paper, he/she would be midway between grade B and the grade A thresholds. Student B's mark is equivalent to a grade E. However, with a little care this score could have been pushed up by 3 or 4 marks, which, if maintained throughout a paper, could have a dramatic effect on the final grade.**

Question 12 Moles; choice of apparatus

Time allocation: 13–15 minutes

A student prepared magnesium chloride, $MgCl_2$, by adding 8.43 g of magnesium carbonate to 2.00 mol dm^{-3} hydrochloric acid.

$$MgCO_3(s) + 2HCl(aq) \rightarrow MgCl_2(aq) + H_2O(l) + CO_2(g)$$

(a) State what the student would see during this reaction. (2 marks)

(b) (i) What amount, in moles, of $MgCO_3$ was used in the experiment? (2 marks)

(ii) Calculate the volume of $2.00 \, mol \, dm^{-3}$ hydrochloric acid needed to react completely with this amount of magnesium carbonate. (2 marks)

(iii) Calculate the volume of CO_2 gas that would be produced at RTP. (2 marks)

(c) When 0.50 g of the carbonates of magnesium, calcium and barium are heated, they decompose and produce an oxide and carbon dioxide.

(i) Write an equation, including state symbols, for the decomposition of one of these carbonates. (2 marks)

(ii) Sketch and label an apparatus that could be used to measure the volume of carbon dioxide evolved. Identify a source of error in your experiment and suggest how the error could be rectified. (4 marks)

(iii) Explain why each carbonate produces a different amount of carbon dioxide. (2 marks)

Total: 16 marks

ⓔ The command word 'calculate' used in b(ii) and (iii) indicates a numerical approach. If more than 1 mark is allocated it is essential to show your working, because any errors in the calculation will be marked consequentially. Incorrect answers may score some marks but only if you show your working. In part (c) of this question practical skills are tested within the written exam.

The marks allocated to each section indicate what is expected. For example, in part (a) there are 2 marks, so two observations are required. In (c)(ii) there are 4 marks, which indicates 1 mark for the diagram, 1 mark for the labels and 2 marks for identifying a possible error and suggesting a remedy for that error — 4 marks = four things wanted.

Student A

(a) Bubbles of carbon dioxide are given off.

Student B

(a) The white solid fizzes and a colourless solution is formed.

ⓔ Student A scores 1 mark but student B scores both marks. The clue is in the question. By looking carefully at the state symbols, you should be able to deduce what you would see. Student A has ignored the fact that 2 marks have been allocated so two observations are required.

Student A

(b) (i) 0.100 mol

Student B

(b) (i) $M_r = 24.3 + 12.0 + 48.0 = 84.3$

   $8.43/84.3 = 0.100\,mol$

ⓔ Each student gains 2 marks. However, student B's technique is better than that of student A. It is always advisable to show the working in calculations. If a mistake is made, some marks can still be awarded for the method. If you make a mistake but show no working, the examiner cannot award any marks.

Student A

(b) (ii) 100

Student B

(b) (ii) moles of HCl = 0.100

   $0.100 = cV$

   $\dfrac{0.100}{2.00} = V = 0.05\,dm^3$

ⓔ Student A has the correct value but scores only 1 mark because there are no units. Student B has taken the reacting ratios to be 1:1 rather than 1:2. However, because he/she has shown some working, 1 mark is awarded for the correct use of $n = cV$.

Student A

(b) (iii) $2400\,cm^3$

Student B

(b) (iii) moles of CO_2 = $0.100 \times 24 = 2.4\,dm^3$

ⓔ Each student scores 2 marks, but student A runs the risk of losing marks by not showing any working.

Student A

(c) (i) $MgCO_3(s) \rightarrow MgO(s) + CO_2(g)$

Student B

(c) (i) $CaCO_3(s) \rightarrow CaO(s) + CO_2(g)$

ⓔ Each student scores 2 marks.

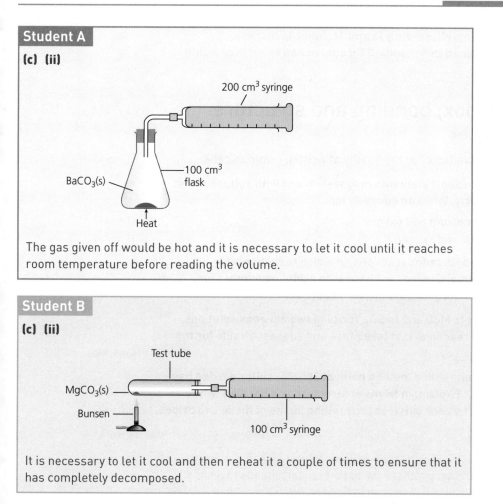

Student A

(c) (ii)

200 cm³ syringe

100 cm³ flask

BaCO₃(s)

Heat

The gas given off would be hot and it is necessary to let it cool until it reaches room temperature before reading the volume.

Student B

(c) (ii)

Test tube

MgCO₃(s)

Bunsen

100 cm³ syringe

It is necessary to let it cool and then reheat it a couple of times to ensure that it has completely decomposed.

ℯ Both students have given good answers. Student A would score all 4 marks even though the apparatus selected is a little large. Student B has clearly labelled the syringe as 100 cm³ but 0.5 g MgCO₃(s) would produce about 142 cm³ of CO₂(g), so student B would score only 3 of the 4 marks.

Student A

(c) (iii) The mass of each carbonate is the same so the moles of each will be different.

Student B

(c) (iii) It depends on the number of moles not the mass. The mole ratio of carbonate to carbon dioxide is 1:1.

ℯ Both students have given good, well thought out answers. Student A scores 1 mark but loses the second mark because they haven't related the moles of carbonate to the moles of CO₂(g). Student B's answer is brief but to the point and gets both marks.

ⓔ Both students have done well, scoring 13 and 14 out of 16 marks, respectively. If they maintained this standard throughout an exam they would both get a grade A.

Question 13 Redox; bonding and structure

Time allocation: 11–13 minutes

In this question, 1 mark is available for the quality of written communication. (1 mark)

(a) A chemist reacts oxygen separately with magnesium and with sulfur to form MgO and SO_2, respectively. Write an equation for:
   (i) the reaction of magnesium and oxygen (1 mark)
   (ii) the reaction of sulfur and oxygen (1 mark)

(b) The reactions in (a) are both redox reactions, in which reduction and oxidation take place. Explain, using the changes in oxidation number for sulfur, whether sulfur undergoes oxidation or reduction. (2 marks)

(c) The chemist adds water to MgO and to SO_2, forming two aqueous solutions. Write equations for the reactions that take place and suggest a value for the pH of each solution. (4 marks)

(d) Magnesium oxide is a solid with a melting point of 2852°C; sulfur dioxide has a melting point of –73°C. Explain, in terms of structure and bonding, why there is such a large difference between the melting points of these two oxides. (5 marks)

Total: 14 marks

ⓔ The bulk of the marks in this question are in parts (c) and (d) and it is worth working out a strategy for answering these parts before starting. Part (c) has 4 marks and requires two equations (1 mark each) and two pH values (1 mark each). Part (d) has 5 marks and requires an explanation of the structure (1 mark) and the bonding (1 mark) of both MgO and SO_2. The fifth mark is for an explanation of the difference in melting points.

Student A
(a) (i) $Mg(s) + \frac{1}{2}O_2(g) \rightarrow MgO(s)$

Student B
(a) (i) $2Mg + O_2 \rightarrow 2MgO$

ⓔ Both students score the mark.

Student A
(a) (ii) $S + O_2(g) \rightarrow SO_2(s)$

> **Student B**
>
> **(a) (ii)** $S + O_2 \rightarrow SO_2$

e Student A could have lost a mark by missing out the state symbol for sulfur, but state symbols are not required here. However, if you do include them, the examiner may penalise you if they are wrong or partly omitted. The best advice is that unless you are asked for state symbols, don't include them in your answer. Both students score the mark.

> **Student A**
>
> **(b)** Initial oxidation state of sulfur is 0 and the final oxidation state of sulfur is +4. Therefore, sulfur has undergone oxidation.

> **Student B**
>
> **(b)** The oxidation state of sulfur changes from 0 to 4.

e Student A scores both marks but student B only scores 1 because they have not stated that the sulfur has been oxidised. Oxidation number has a sign as well as a value. The minus sign should always be included for negative oxidation numbers. If the oxidation number is positive, the + sign should be written, but if there is no sign, the examiner will assume the number to be positive.

> **Student A**
>
> **(c)** $MgO + H_2O \rightarrow Mg(OH)_2$ pH of the aqueous solution = 11
>
> $SO_2 + H_2O \rightarrow H_2SO_3$ pH of the aqueous solution = 2

> **Student B**
>
> **(c)** $MgO + H_2O \rightarrow MgOH_2$ pH = 9
>
> $SO_2 + 2H_2O \rightarrow H_2SO_4 + H_2$ pH = 5

e Student A scores all 4 marks. The pH depends on the concentration, so a value between 8 and 13 is acceptable for $Mg(OH)_2$, as is a value between 1 and 6 for H_2SO_3. Student B gets 2 marks for the pH predictions but loses both equation marks. The first equation is almost correct, but the formula for magnesium hydroxide must have brackets around the OH. The second equation is incorrect, because the student tries to form H_2SO_4 instead of H_2SO_3.

> **Student A**
>
> **(d)** MgO has a high melting point because it has a giant ionic lattice. SO_2 has a low melting point because it is a covalent molecule with weak intermolecular forces.

Student B

(d) MgO has strong ionic bonds throughout the lattice, which need a lot of energy to break them. SO_2 has covalent bonds that are weak and, therefore, little energy is required to melt it.

ⓔ The mark scheme is shown below:

- MgO:
  - bonding — ionic ✓
  - structure — giant lattice with strong bonds throughout ✓
- SO_2:
  - bonding — covalent ✓
  - structure — simple molecular with weak intermolecular forces ✓
- MgO has a high melting point because it requires a lot of energy to break the strong bonds within the lattice. SO_2 has a low melting point because only a small amount of energy is needed to break the weak intermolecular forces between the molecules. ✓

ⓔ Student A scores 4 marks for correctly describing the bonding and structure in each compound, but they have not related this to the size of the melting point. Student B also scores 4 marks. A mark is lost for failing to mention that the structure in SO_2 is simple molecular.

ⓔ **Both students have answered the structured parts well, with student A scoring the maximum 8 marks and student B scoring 5. The free-response question proves difficult for many students, and here both have lost marks. When attempting to answer free-response questions it is essential that you devise a plan based on the information given in the question. On the free-response question both students have scored 4 marks. Overall, student A scores 12 marks and student B scores 9 marks.**

Question 14 Electronegativity; hydrogen bonding

Time allocation: 9–10 minutes

(a) State what is meant by the term 'electronegativity'. (2 marks)

(b) (i) Draw a diagram to show hydrogen bonding between two molecules of water. Your diagram must include dipoles and lone pairs of electrons. (4 marks)

(ii) State the bond angle in a water molecule. (1 mark)

(c) State and explain two properties of ice that are a direct result of hydrogen bonding. (4 marks)

Total: 11 marks

ⓔ When you are asked to draw a diagram a substantial amount of care is required. In (b)(i) there are 4 marks so a mark will be allocated to each of four points: dipoles, shape, position of the H-bond and the involvement of the lone pairs on the O.

Student A

(a) Attraction for the shared pair of electrons within a covalent bond.

Student B

(a) Different atoms have different attractions for electrons.

ℯ Student A gains both marks. Student B scores only 1 mark, for a partly correct answer that does not mention the covalent bond.

Student A

(b) (i)

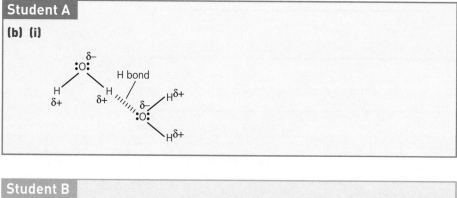

Student B

(b) (i)

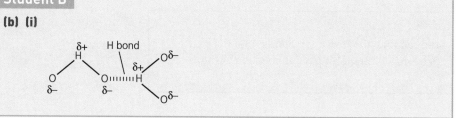

ℯ The four marking points are:

• correct dipoles with $O^{\delta-}$ and $H^{\delta+}$ ✓

• water drawn as non-linear ✓

• the hydrogen bond between a hydrogen atom in one water molecule and the oxygen atom in an adjacent water molecule ✓

• the involvement of the lone pair of electrons on oxygen and the linear shape of $O-H\cdots O$ ✓

Student A gives the perfect answer and earns all 4 marks. To obtain full marks, the hydrogen bond must be drawn carefully so that the sketch shows clearly the involvement of a lone pair of electrons on the oxygen atom, which many students fail to do.

Questions & Answers

A significant minority of students draw water as HO_2, not H_2O. Should student B lose all 4 marks? The dipoles are correct, the shape is correct and a hydrogen bond has been drawn between the hydrogen in one water molecule and the oxygen in another water molecule. However, there is no indication of the involvement of the lone pair of electrons, and the H—O·····H is not drawn as linear. Student B scores 1 mark (maximum 2) but could have gained 3 marks by drawing the structure of water correctly. It is easy to lose marks through carelessness.

Student A

(b) (ii) Approximately 104°

Student B

(b) (ii) 104.5°

ⓔ Both students score the mark for this simple recall question.

Student A

(c) Ice floats on water because air is trapped between the water molecules. Ice has a higher melting point than expected because of the hydrogen bonds.

Student B

(c) Ice is less dense than water because the hydrogen bonds hold the molecules further apart. Ice has a high melting point because of the hydrogen bonds.

ⓔ Each student scores 3 marks, for different reasons. Student A gets 2 marks for the two properties of ice, but the explanation for the fact that ice floats on water is incorrect. Student B almost gains 4 marks. However, the statement that 'ice has a high melting point' should be qualified by writing 'a higher melting point than expected' in order to earn the mark.

ⓔ **Students A and B seem to have similar ability, but the outcome does not reflect this, with student A scoring more marks than student B. Losing marks when you know the chemistry is usually down to either examination technique or carelessness. In this question, student A scores 10 marks out of 11, whereas student B scores only 6 marks. Student A could achieve a grade A; student B's scores fluctuate between grade A and grade D. It is useful to look at student B's responses and identify the sort of errors/slips made. If you can recognise the mistakes made by others, you may avoid making them yourself.**

Question 15 Planning an experiment; interpretation of results

Time allocation: 12 minutes

A student was provided with a sample of calcium carbonate that was known to be impure. He was asked to design an experiment to calculate the percentage purity of the sample. He decided to react a known mass of the sample with excess hydrochloric acid and collect the carbon dioxide evolved by the displacement of water.

He used a measuring cylinder to measured 100 cm^3 of 0.50 mol dm^{-3} HCl(aq) and set up the apparatus as shown in Figure 1.

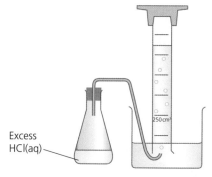

Excess
HCl(aq)

Figure 1

The student then weighed 1.00 g of CaCO$_3$(s), removed the bung, added the CaCO$_3$(s) and quickly replaced the bung. He measured the volume of water displaced and used his results to estimate the purity of the sample.

(a) **(i)** Identify two possible errors in the method.   (2 marks)

   (ii) Identify two possible measurement errors.   (2 marks)

   (iii) Suggest an improvement for one of the errors that you identified
   in the method.   (1 mark)

(b) Calculate how the minimum volume of 0.5 mol m^{-3} HCl(aq) would be required
to ensure that it was in excess.   (4 marks)

(c) Why is it necessary to have an excess of HCl(aq)?   (1 mark)

(d) A second student carried out the same experiment but decided to collect the
carbon dioxide in a syringe rather than by displacement of water. She found
that when 0.50 g of impure CaCO$_3$(s) reacted with excess HCl(aq), 96 cm^3 of
CO$_2$(g) was collected in the syringe. Calculate the percentage purity of the
sample of CaCO$_3$(s).   (4 marks)

Total: 14 marks

e The command word 'identify' requires a simple statement and doesn't need
an explanation to justify the identification. If an explanation were required, the
command would be to 'identify' and 'explain your reasoning'.

Student A

(a) (i) CO_2 is soluble in water and $CO_2(g)$ would escape before the bung could be replaced.

Student B

(a) (i) CO_2 dissolves in water, the HCl(aq) should be measured more accurately using a burette.

ⓔ Student A gets both marks but student B scores only 1 mark. It is true that measuring cylinders are not very accurate but the HCl(aq) is in excess so it doesn't need to be measured accurately.

Student A

(a) (ii) Any two from measuring the HCl(aq), weighing the $CaCO_3(s)$ or measuring the volume of $CO_2(g)$.

Student B

(a) (ii) The volume of HCl used and the volume of $CO_2(g)$ collected.

ⓔ All measurements have in-built errors. However, the volume of HCl(aq) used is not relevant as long as it is in excess. Unfortunately, student A is being a bit too clever and loses 1 mark. Student B also loses the mark for the HCl.

Student A

(a) (iii) Use lemonade instead of water.

Student B

(a) (iii) Get a friend to help you add the $CaCO_3(s)$ so that the bung can be replaced quickly.

ⓔ Student A would get the mark because lemonade, like all fizzy drinks, is carbonated and already contains dissolved $CO_2(g)$. Although student B's answer seems reasonable, it is unlikely to score the mark because $CO_2(g)$ would still escape.

Student A

(b) $40\,cm^3$

Student B

(b) $CaCO_3(s) + 2HCl(aq) \rightarrow CaCl_2(aq) + H_2O(l) + CO_2(g)$

mols of $CaCO_3(s) = \dfrac{1}{100.1} = 0.0099$

Therefore need 0.1199 mols of HCl(aq)

volume of HCl $= \dfrac{0.0199}{0.5} = 0.03996\,dm^3 = 39.96\,cm^3$

ℯ Both students get full marks but student A is taking an enormous risk. By just quoting the answer student A either gets full marks or zero marks. Student B's approach is much more sensible because each step in the calculation can be tracked and any errors identified. If an error is made early in a calculation, marks can still be awarded as the 'error is carried forward' and marked as 'ecf ✓'.

Student A

(c) To make sure all the $CaCO_3$ has reacted.

Student B

(c) Only the impurities in the calcium carbonate will be left.

ℯ Both students get the mark.

Student A

(d) Moles of $CO_2 = 4 \times 10^{-3}$, which is the same as the moles of pure $CaCO_3$.

The mass of pure $CaCO_3 = 4 \times 10^{-3} \times 100.1 = 0.4004\,g$

The % purity $= \dfrac{0.4004}{0.5} \times 100 = 80.1\%$

Student B

(d) moles of impure $CaCO_3$ used $= \dfrac{0.5}{100.1} = 0.004995\,mol$

moles of CO_2 actually collected $= \dfrac{96}{24000} = 0.004\,mol$

% purity of $CaCO_3 = \dfrac{0.004}{0.004995} \times 100 = 80.1$

ℯ Both students score all 4 marks. Student A demonstrates good exam technique and shows all the working. Student B calculates the percentage in an unorthodox way but gets the correct value and is therefore awarded all the marks.

ⓔ Overall, student A scores 13 out of 14 marks, while student B gains 11 marks. Both have answered this question well, but student B might be disappointed with 11 out of 14, which equates to a grade B answer.

Question 16 Bonding and related properties

Time allocation: 20 minutes

Chemists have developed models for bonding and structure, which are used to explain different properties.

(a) Ammonia, NH_3, is a covalent compound.

    (i)   Explain what is meant by a covalent bond.   (1 mark)

    (ii) Draw a 'dot-and-cross' diagram of NH_3.   (1 mark)

    (iii) State, and explain, the shape of the ammonia molecule.   (5 marks)

    (iv) State one property of NH_3 that is typical of covalent substances.   (1 mark)

(b) Ammonia reacts with an acid to form an ammonium ion. Write an ionic equation for this reaction and predict the shape of and bond angle in the ammonium ion.   (3 marks)

(c) An amide ion, NH_2^-, can also be formed. Draw a 'dot-and-cross' diagram of an amide ion, suggest its shape and state the likely bond angles in the ion.   (3 marks)

(d) Ammonium compounds such as ammonium sulfate, $(NH_4)_2SO_4$, can be used as fertilisers.

    (i)   Write a balanced equation to show how ammonium sulfate could be formed by the reaction between aqueous ammonium hydroxide and sulfuric acid.   (1 mark)

    (ii) Ammonium sulfate is an example of a salt formed when an acid is neutralised by a base. Explain what is meant by the term salt.   (1 mark)

    (iii) When ammonium sulfate is formed by reacting ammonia with sulfuric acid, explain why ammonia is acting as a base.   (1 mark)

(e) Ammonium sulfate is an ionic substance.

    (i)   Give the formula of the ions present and deduce the oxidation states of each element in each ion.   (4 marks)

    (ii) State two properties of ammonium sulfate that are typical of ionic substances.   (2 marks)

Total: 23 marks

ⓔ This is a straightforward question and the bulk of the marks can be scored by careful drawing of dot-and-cross diagrams and application of the electron pair repulsion theory. The amide ion, NH_2^-, is not a standard ion and you are expected to use your knowledge and understanding to deduce its shape.

> **Student A**
>
> (a) (i) A covalent bond is a shared pair of electrons.

Student B

(a) (i) A covalent bond is formed when two electrons are shared between two atoms and each atom provides one of the shared electrons.

e Both answers are essentially correct but neither offers the full definition given in the specification section 2.2.2d, which states that a 'covalent bond is the electrostatic attraction between a shared pair of electrons and the nuclei of the two bonded atoms', so it is unlikely that either student will score the mark.

Student A

(a) (ii)

Student B

(a) (ii)

e Both students score the mark.

Student A

(a) (iii) NH_3 is shaped pyramidal, bond angle 107°.

Student B

(a) (iii) Shape = pyramidal. Bond angle = 107°. The shape depends on the number of electron pairs, which repel each other. Lone pairs repel more than bonded pairs. The central N has four pairs (three bonded and one lone) so the bond angle is compressed from 109.5° to about 107°.

e Student A clearly understands the chemistry and scores 2 marks for the shape and the angle, but they also appear to be in a rush and haven't answered the most important part of the question that asks for an explanation, which automatically loses 3 marks. Student B has provided the perfect answer and scores all 5 marks.

Student A

(a) (iv) It is a gas.

> ### Student B
>
> **(a) (iv)** It has a low boiling point because it is a simple molecule that only has weak intermolecular forces between the molecules, so very little energy is needed to break the intermolecular forces, hence a low boiling point.

ⓔ Both score the mark but this time student B is guilty of not reading the question, which requires a simple statement and not an explanation. Try to develop your examination technique and use the information in the question. If there is only 1 mark then the answer required will likely be brief and straightforward.

> ### Student A
>
> **(b)** $NH_3 + H^+ \rightarrow NH_4^+$
>
> Ammonium ion is tetrahedral and bond angles are 109° 28′.

> ### Student B
>
> **(b)** $NH_3 + H^+ \rightarrow NH_4^+$
>
> 109.5°, the shape is a tetrahedron.

ⓔ Both students give good answers and both score all 3 marks.

> ### Student A
>
> **(c)**
>
>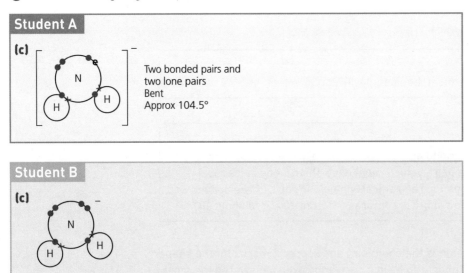
>
> Two bonded pairs and two lone pairs
> Bent
> Approx 104.5°

> ### Student B
>
> **(c)**
>
> Shape is angular with angles ~104.5°

e Both students give good answers and student A scores all 3 marks. Student B loses a mark for the diagram because there appears to be six identical electrons (all drawn as red dots) around the central N, which implies that the N has six electrons in its outer shell, whereas N is in group 5 and therefore has only five electrons in its outer shell.

Student A

(d) (i) $2NH_4OH + H_2SO_4 \rightarrow (NH_4)_2SO_4 + 2H_2O$

Student B

(d) (i) $2NH_3(aq) + H_2SO_4(aq) \rightarrow (NH_4)_2SO_4(aq)$

---

e Student A has answered the question but student B has used the knowledge that $NH_3(aq)$ is equivalent to ammonium hydroxide in that the ammonia reacts with the water to form ammonium hydroxide, which would then react with sulfuric acid to produce ammonium sulfate and water. Student B would also get the mark.

Student A

(d) (ii) A salt is formed when the acidic proton (H^+) in an acid is replaced by a cation.

Student B

(d) (ii) A salt is the product formed when an acid is neutralised by a base.

---

e Student A gets the mark since a cation is a positive ion, such as a metal ion or an ammonium ion. Student B doesn't score the mark, partly because the answer simply repeats what was in the question, but partly because when a salt is formed in a neutralisation reaction, other products are formed as well.

Student A

(d) (iii) It accepts a proton.

Student B

(d) (iii) It reacts with an acid.

---

e Student A gets the mark by stating the classic definition of a base. Student B doesn't score the mark because acids can react with other substances (such as silver nitrate: $HCl(aq) + AgNO_3(aq) \rightarrow AgCl(s) + HNO_3(aq)$) and not necessarily behave as an acid in that reaction and once again he/she is simply repeating what is in the question.

Student A

(e) (i) The ammonium ion and the sulfate ion NH_4^+. The N is –3 and H is +1.

(ii) Ionic compounds have high boiling points and dissolve in water.

Student B

(e) (i) and (ii)

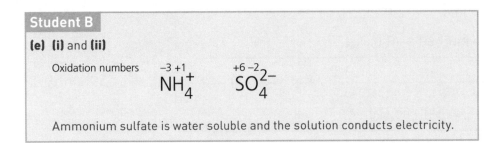

Oxidation numbers $\overset{-3\ +1}{NH_4^+}$ $\overset{+6\ -2}{SO_4^{2-}}$

Ammonium sulfate is water soluble and the solution conducts electricity.

🅮 Student B gives the perfect answer and scores all 6 marks. Student A seems to have forgotten about the sulfate ion and in part (i) scores just 1 mark for the NH_4^+. Student A also scores 1 mark for the oxidation numbers of the elements in the ammonium ion and a further 2 marks in part (ii).

🅮 **Overall, student A scores 18 out of 23 and student B scores 20.**

Question 17 Acids, bases and ionic equations

Time allocation: 10 minutes

(a) In the presence of oxygen, magnesium powder burns with a bright white light, to produce magnesium oxide solid.

   (i) Write an equation, including state symbols, for this reaction.   2 marks)

   (ii) Draw a dot-and-cross diagram of magnesium oxide.   (2 marks)

(b) Magnesium oxide is a base and reacts with acids.

   (i) Explain what is meant by 'a base'.   (1 mark)

   (ii) Write an equation for the reaction of magnesium oxide with nitric acid.   (1 mark)

   (iii) Explain whether or not this is a redox reaction.   (1 mark)

(c) A student reacted identical masses of magnesium oxide with equal volumes of sulfuric acid and with ethanoic acid. Both acids had the same concentration.

   (i) Write an ionic equation for the reaction between solid magnesium oxide and sulfuric acid.   (2 marks)

   (ii) State the formula of the salt formed when magnesium oxide reacts with ethanoic acid.   (1 mark)

   (iii) State what the student would have observed in both reactions.   (1 mark)

   (iv) Suggest one difference that the student would have observed. Explain that difference.   (2 marks)

      Total: 13 marks

ⓔ In order to do well in this question you need to know the formulae of the common acids that are listed in section 2.1.4(a) of the specification and you must be able to write equations and deduce formulae.

Student A

(a) (i) $Mg(s) + \frac{1}{2}O_2(g) \rightarrow MgO(s)$

Student B

(a) (i) $Mg(s) + O(g) \rightarrow MgO(s)$

ⓔ Student A gets both marks. student B scores just 1 mark for the correct state symbols, even though the formula of oxygen is incorrect.

Student A

(a) (ii)

Student B

(a) (ii)

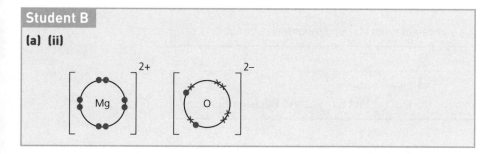

ⓔ Student A has done what many students do and incorrectly drawn MgO as covalent when it is ionic. Student B gets both marks. It is not necessary to draw the Mg^{2+} ion showing eight electrons in the outer shell — full marks could be achieved by simply drawing:

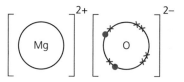

Student A

(b) (i) A base can accept a proton.

> **Student B**
>
> **(b) (i)** Acids donate protons and bases accept them.

ⓔ Both students score the mark.

> **Student A**
>
> **(b) (ii)** $MgO + 2HNO_3 \rightarrow Mg(NO_3)_2 + H_2O$

> **Student B**
>
> **(b) (ii)** ~~$MgO + HNO_3 \rightarrow MgNO_3 + H_2O$~~
>
> $\quad\quad MgO + H_2NO_3 \rightarrow MgNO_3 + H_2O$

ⓔ Student A gets the mark, has clearly taken time to learn the formula of the common acids and knows how to work out formulae of common salts. This is not the case with student B, who initially had the formula of nitric acid correct, but when the equation didn't balance simply changed the formula of nitric acid to make it balance. Student B should have corrected the formula of the salt magnesium nitrate.

> **Student A**
>
> **(b) (iii)** It isn't a redox reaction because no oxidation states have changed.

> **Student B**
>
> **(b) (iii)** It is a redox reaction because Mg has lost oxygen and H has gained it.

ⓔ The best way to decide whether or not a reaction is a redox reaction is by considering the oxidation states in the equation. It is worth writing out:

Oxidation numbers:
$$\begin{array}{cccc} +2\,-2 & +1\,+5\,-6 & +2\,+5\,-2 & +1\,-2 \\ MgO & +\ 2HNO_3 & \rightarrow\ Mg(NO_3)_2 & +\ H_2O \end{array}$$

Student A scores the mark but student B gets no marks.

> **Student A**
>
> **(c) (i)** $MgO(s) + 2H^+(aq) \rightarrow Mg^{2+}(aq) + H_2O(l)$

> **Student B**
>
> **(c) (i)** $O^{2-} + 2H^+ \rightarrow H_2O$

e When writing ionic equations it is essential to include state symbols and to understand that ions only exist as free ions when they are in aqueous solution. MgO is an ionic solid but it doesn't exist as free Mg^{2+} and free O^{2-} ions. Student A has done well and gets both marks. Student B displays some understanding but scores no marks.

Student A

(c) (ii) $Mg^{2+}(^-OOCCH_3)_2$

Student B

(c) (ii) CH_3COOMg

e Student A has done well and worked out the correct formula. It helps to show the charge of each ion and remember that the charges must balance and the compound must be neutral. Student B has made a good attempt and has remembered the formula of ethanoic acid as CH_3COOH but hasn't taken account of the charges on each ion, so doesn't get the mark.

Student A

(c) (iii) Both reactions would get hot.

Student B

(c) (iii) The MgO solid reacts and dissolves and a gas is given off.

e The correct observation is that the MgO solid reacts with the acid and forms a colourless solution of magnesium sulfate. Student B has done what many students do whenever they are asked to describe what they would see, which is to instinctively write down 'gas given off' or 'bubbles'. However, this only happens when one of the products is a gas, usually either $H_2(g)$ or $CO_2(g)$. Student B initially stated the correct observation but then lost the mark by going on to say that a gas was evolved. Student A might get the mark as the reactions are exothermic but the examiners would have decided whether or not to allow 'heat' as an observation.

Student A

(c) (iv) The reaction with sulfuric acid would be faster because it is a stronger acid than ethanoic acid.

Student B

(c) (iv) The bubbles of gas would be faster with H_2SO_4 than with CH_3COOH because CH_3COOH is a weak acid.

e Both students have done well. Student A scores both marks, as does student B. Student B gets both marks even though there are no 'bubbles of gas'. This incorrect statement was penalised in the previous part of the question and would not be penalised again.

e Overall, student A has done really well and scored 11 marks out of 13. The only marks lost by student A were in the dot-and-cross diagram of MgO. These marks would be regarded as easy marks and shouldn't be lost by a good candidate. Student B has done less well and scores 6 out of 13 marks, which equates to a grade E answer. Student B clearly has a problem with formulae and with writing balanced equations. It is important to recognise your own weaknesses and to work on improving them.

Question 18 Practical skills in chemistry

Time allocation: 10 minutes

A student was asked to design and carry out an experiment to confirm that the value of x in the formula of barium chloride crystals, $BaCl_2.xH_2O$, is 2: that is, the formula is $BaCl_2.2H_2O$. The student devised the method as follows.

The student weighed a crucible and recorded the mass in a table. He also recorded the error range of the balance. He then added the $BaCl_2.xH_2O$ crystals to the crucible, re-weighed the crucible with the crystals and recorded this result in a table.

Next, the student placed the crucible containing the crystals on a pipe-clay triangle suspended on a tripod. He used a Bunsen burner to heat the crucible, using a gentle heat at first, and then using a hot flame, for between 3 and 4 minutes.

After heating, the crucible and its contents were allowed to cool. The student then weighed them again and recorded the results in the table. He then re-heated the crucible and its contents for a further 2 minutes, cooled and re-weighed it.

The results are shown in Table 3.

Measurement	Mass in g	Error range of balance
Mass of crucible	14.94	+/– 0.01 g
Mass of crucible + $BaCl_2$ crystals	17.59	+/– 0.01 g
Mass of crucible + $BaCl_2$ after first heating	17.34	+/– 0.01 g
Mass of crucible + $BaCl_2$ after second heating	17.29	+/– 0.01 g

Table 3

(a) Use the student's results to calculate:

    (i) the mass of $BaCl_2$ crystals before heating and the mass of $BaCl_2$ after heating   (2 marks)

    (ii) the mass of water in the $BaCl_2$ crystals   (1 mark)

    (iii) the amount in moles of $BaCl_2$ and H_2O (3 marks)

    (iv) the value of x in $BaCl_2 \cdot xH_2O$ (2 marks)

(b) Calculate the percentage error:

   (i) when weighing the crucible   (1 mark)

   (ii) in the mass of water   (1 mark)

(c) Identify a fault, if any, in the procedure used by the student and suggest an improvement.   (1 mark)

Total: 11 marks

e Practical skills will be assessed by your teachers but will also be examined in written papers. It is essential that you are able to plan, implement, analyse and evaluate a range of practical procedures.

> **Student A**
>
> **(a) (i)** 2.65 g and then 2.35 g

> **Student B**
>
> **(a) (i)** 17.59 − 14.94 = 2.65 g and 17.29 − 14.94 = 2.35 g

e Both students score 2 marks.

> **Student A**
>
> **(a) (ii)** 0.30 g

> **Student B**
>
> **(a) (ii)** 17.59 − 17.29 = 0.3 g

e Student A scores the mark, whereas student B would probably lose the mark because all the weight readings have been recorded to two decimal places and it would be expected that the mass of the water would also be written to two decimal places. The examiner has deliberately picked the numbers such that the second decimal place is a '0' and student B has fallen into the trap of not writing it in.

> **Student A**
>
>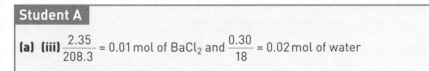
> **(a) (iii)** $\dfrac{2.35}{208.3}$ = 0.01 mol of $BaCl_2$ and $\dfrac{0.30}{18}$ = 0.02 mol of water

> **Student B**
>
>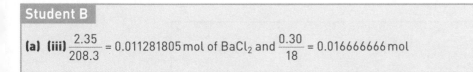
> **(a) (iii)** $\dfrac{2.35}{208.3}$ = 0.011281805 mol of $BaCl_2$ and $\dfrac{0.30}{18}$ = 0.016666666 mol

e The marks awarded in this calculation are for:

molar mass/g mol^{-1} of $BaCl_2$ = 208.3 — both students score the mark.

moles of $BaCl_2$ — student A has rounded the number in the middle of a calculation and loses the mark, but student B has copied the calculator value, which would get the mark.

moles of H_2O — student A has rounded again but wouldn't be penalised a second time for the same error, so both students would get the mark.

Student A

(a) (iv) ratio $BaCl_2 : H_2O$ is 0.1 : 0.2, which is 1 : 2

Hence the value of x is 2.

Student B

(a) (iv) ratio $BaCl_2 : H_2O$ is 0.011281805 : 0.016666666

which is 1 : 0.016666666/0.011281805 = 1.477304917

Hence the value of x is 1.

ⓔ Student A has correctly used his answers from (a)(iii) and therefore scores both marks, as does student B. The value obtained by student B leaves considerable doubt as to the actual value of x — which is often the case with experiments. It is rare for an experiment to give perfect answers.

Student A

(b) (i) $\dfrac{0.01}{14.94} \times 100 = 0.067\%$

Student B

(b) (i) 0.067%

ⓔ Both students get the mark but this time student A shows better exam technique.

Student A

(b) (ii) $\dfrac{0.02}{0.30} \times 100 = 6.7\%$

Student B

(b) (ii) $\dfrac{0.01}{0.30} \times 100 = 3.3\%$

---

e Student A has done well and spotted the catch. To get the mass of the water two weighings were required, so the error was doubled.

> **Student A**
>
> **(c)** There are no faults — the calculation gave the perfect answer.

> **Student B**
>
> **(c)** The student should have heated the crucible and crystals again until the mass stopped changing.

e Student B gets the mark but student A does not, even though their calculated value to (a)(iv) does give the exact value. In an experiment like this it is essential to heat, cool and weigh several times until you are certain that a constant mass has been achieved.

e **Overall, both students score 9 out of 11 marks, which is grade A standard. With a little care and thought student A could have got full marks. In any calculation it is essential *not* to round during the calculation. All figures should be kept in the calculator until the end of the calculation and only then should you round to an appropriate number of significant figures.**

Extended response questions

Questions in this section are relevant to AS Component 2 and to A-level Component 3. Question 21 is for A-level only.

Question 19 Bonding

Time allocation: 12–14 minutes

In this question, 1 mark is available for the quality of written communication. (1 mark)

Chlorine reacts with sodium to form sodium chloride.

(a) **Describe the bonding in sodium, chlorine and sodium chloride.** (7 marks)

(b) **Relate the physical properties of chlorine and sodium chloride to their structures and bonding.** (8 marks)

Total: 16 marks

e The command word 'describe' requires a statement usually supported by appropriate diagrams. 'Relate' requires a detailed description of how properties are dependent on their structure and bonding.

Questions & Answers

Student A

(a) Sodium contains a giant metallic bond. This is formed by the close packing of the atoms so that their outer-shell electrons overlap and the electrons are free to move anywhere in the lattice. The mobile electrons are like a 'glue' that holds the lattice together.

Chlorine is a covalent molecule made by sharing electrons.

Sodium chloride forms ionic bonds. An ionic bond is the electrostatic attraction between oppositely charged ions, which results from the transfer of an electron from the sodium to the chlorine.

Student B

(a)

Sodium	Chlorine	Sodium chloride
Metallic bonding	Covalent bonding	Ionic bonding
There is a giant lattice of positive ions surrounded by a sea of mobile electrons	There is a shared pair of electrons in the bond	The sodium transfers an electron to the chlorine and the oppositely charged ions are attracted to each other

ⓔ The mark scheme is shown below:

- sodium — metallic ✓; lattice of positive ions ✓; delocalised/mobile electrons ✓
- chlorine — covalent ✓; sharing electrons ✓
- sodium chloride — ionic ✓; electrostatic attraction between oppositely charged ions ✓

Student A scores 5 of the 7 marks available. In the bonding of sodium, student A has missed out 'positive ions' in the description of the lattice, and the diagram drawn for the structure of sodium chloride contradicts the description in words. Student A explains ionic bonding correctly but then draws a diagram showing shared electrons, making it covalent. Student A's answer is ambiguous — it does not make clear whether NaCl is ionic or covalent. The examiner will not select the correct response for you. If you give two pieces of information that are contradictory, the examiner will always mark the incorrect answer first. Student B scores all 7 marks. In these circumstances, a table is an acceptable way of answering the question.

Student A

(b) Chlorine is a gas because the bonds between the Cl_2 molecules are weak induced dipole–dipole forces. It is a poor conductor because it contains no mobile electrons.

Sodium chloride has a high melting point because there are strong ionic bonds throughout the lattice. It conducts electricity when molten or aqueous but not when solid. This is because when molten or aqueous the electrons are free to move.

Student B

(b)

Substance	Property
Chlorine	Gas; poor conductor; insoluble in water
Sodium chloride	Solid; conducts when molten or aqueous; dissolves in water

---

ⓔ The mark scheme is shown below:

- chlorine — two properties ✓✓; two reasons explaining the properties ✓✓
- sodium chloride — two properties ✓✓; two reasons explaining the properties ✓✓

Students often find free-response questions difficult. However, there are always clues in the question. Here, you are given two substances, chlorine and sodium chloride, and asked to relate their physical properties to their structures and bonding. There are 8 marks available, so it is logical to give two properties for each, together with explanations of each property, that is, eight different points for 8 marks.

Student A scores all 4 marks for chlorine but loses a mark for sodium chloride because sodium chloride conducts electricity by the movement of ions, not electrons.

Student B has again chosen to use a table, which is a valid way to answer the question. However, there are several errors. Three general properties of covalent substances are listed but the third, with respect to chlorine, is incorrect. Chlorine is slightly soluble in water and reacts with water to produce a mixture of HCl and HClO. Student B obtains only 1 mark for the properties, even though two properties are correct (remember the third property is wrong). For sodium chloride, student B lists three properties that are all correct and so gains 2 marks, but has forgotten to give explanations for the properties and automatically loses 4 marks.

ⓔ **Examiners accept that chemists communicate in various ways: equations, diagrams and tables. However, whenever there are marks for quality of written communication, it is advisable to write in sentences, as the examiner is often looking for one or two chemical terms used correctly and in an appropriate context. Student A gains the mark for quality of written communication and scores a total of 13 marks out of 16. Student B loses the mark for quality of written communication because the answer contains no continuous prose. Student B scores 10 marks.**

Question 20 Shapes of molecules

Time allocation: 8–10 minutes

In this question, 1 mark is available for the quality of written communication.   (1 mark)

Electron-pair repulsion theory can be used to predict the shapes of covalent molecules. State what is meant by the term 'electron-pair repulsion theory' and use it to determine the shapes of four molecules of your choice. Choose molecules to illustrate four different shapes. State the bond angle in each shape.   (10 marks)

Total: 11 marks

ⓔ When asked to 'draw' shapes of molecules, three-dimensional diagrams are expected where appropriate, with bonds shown as 'wedges' or as dashed or dotted lines.

Student A

The shape of a molecule depends on the number of electron pairs around the central atom. Each pair repels the others but lone pairs of electrons repel more than bonded pairs.

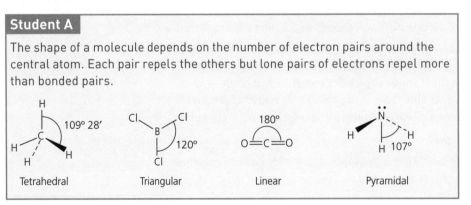

Student B

Electron-pair repulsion theory is a theory that tells us about electron-pair repulsion. It tells us that electrons repel each other and therefore electrons do not pair up because they repel each other.

The shape of molecules depends on the number of bonds in a molecule.

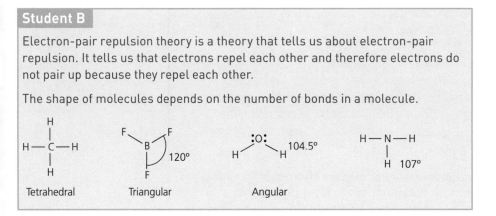

| Tetrahedral | Triangular | Angular | |

(e) Student A has drawn the shapes of molecules in three dimensions and gives an excellent answer, scoring 2 marks for the description of electron-pair repulsion theory and 2 marks each for the four shapes. The quality of written communication mark is awarded because specific chemical terms are used in the correct context with due regard to spelling, punctuation and grammar. Student B has not done enough revision and does not know the straightforward definition of the electron-pair repulsion theory. In what appears to be a rushed answer, the question has been restated. This fails to score. Student B's sketches for methane and ammonia are two-dimensional and are insufficient to gain the marks. It is particularly important to show the three-dimensional shapes of these molecules. This is best done using wedge-shaped bonds. Student B also loses a mark for not quoting the bond angle in methane but does earn the mark for quality of written communication. Student B's score for the four shapes is 5 marks.

(e) **Overall, student A scores the full 11 marks, while student B scores 6. Often, chemists do not write in continuous prose but communicate by symbols, equations and specific chemical terms. It is possible to gain the quality of written communication mark without writing at great length. As little as two consecutive sentences may be regarded as continuous prose.**

Question 21 Unstructured mole calculation

A-level only. Time allocation: 6 minutes

0.134 g of a group 2 metal, M, was added to 100 cm^3 water and 80.2 cm^3 of hydrogen was collected. Identify the metal, M, and calculate the concentration of the resulting solution. Show all of your working. (7 marks)

Questions & Answers

Student A

Group 2 metals react with water to give hydrogen and the hydroxide:

$$M + 2H_2O \rightarrow H_2 + M(OH)_2$$

Volume of H_2 = 80.2

So mols = $\dfrac{80.2}{24000}$ = 3.34 × 10^{-3}

Therefore mols of M = 3.34 × 10^{-3}

Mass of M = 0.134 g, hence relative molecular mass of M = $\dfrac{0.134}{3.34 \times 10^{-3}}$ = 40.09

M is Ca

Concentration of the solution is 3.34 × 10^{-3} mol dm^{-3}

Student B

$$M + 2H_2O \rightarrow M(OH)_2 + H_2$$

Mols of H_2 = 3.34 × 10^{-2} = mols of M

Mols of Be = $\dfrac{0.134}{9}$ = 0.149 mols so its not Be

Mols of Mg = $\dfrac{0.134}{24.3}$ = 5.51 × 10^{-3} so its not Mg

Mols of Ca = $\dfrac{0.134}{40.1}$ = 3.34 × 10^{-3} so it is Ca

---

(e) Open-ended calculations like this are much more likely to appear on A-level papers than on AS papers. There are a number of ways to solve this question and the mark scheme would be open to allow for different approaches.

The 7 marks would be allocated for: correct equation ✓; amount in moles of H_2 ✓; amount in moles of metal M ✓; relative atomic mass of M ✓; correct identity of M ✓; concentration of $M(OH)_2$ taking into account the volume of water used ✓; units of concentration ✓.

Student A scores marks 1 to 5, as listed, but might be penalised 1 mark for describing the relative atomic mass as the relative *molecular* mass. Student A scores the mark for the units of concentration, but has not scored the mark for the concentration of $M(OH)_2$.

Student B has taken a very different approach and correctly identified Ca by a series of trial and error calculations and would be awarded marks 1 to 5. Student B seems to have forgotten about calculating the concentration of the solution formed.

Knowledge check answers

1. **(a)** flammable
   (b) hazardous to the environment
   (c) explosive substance
   (d) corrosive
   (e) toxic
   (f) oxidising substances
   (g) irritant
   (h) radioactive

2. 1F, 2H, 3D, 4A, 5C, 6I, 7B, 8E, 9G

3. **measuring cylinder** — measure approximate volume of liquid; **pipette** — measure a set amount of liquid; **burette** — measure any volume of liquid accurately; **condenser** — liquefy (condense) volatile liquids; **separating funnel** — separate immiscible liquids; **gas collection** — collect water-insoluble gas.

4. **(a) (i)** 2%
       (ii) 0.2%
       (iii) 0.2%
   (b) (i) 4%
       (ii) 2% (but the error could be doubled if initial and final weighings are taken)
   (c) (i) 16.7%
       (ii) 3.3% (two readings have to be made so the error is doubled)

5. **(a)** 735
   (b) 6.98×10^3
   (c) 0.000346
   (d) (i) $0.41\,g\,cm^{-3}$
       (ii) $0.406\,g\,cm^{-3}$
       (iii) $0.120\,g\,cm^{-3}$

6. Remove: 25.0 because it is the rough initial titration and result is not recorded to 2 decimal places. Remove 26.50, not concordant with the other values. Average titre value is $(25.40 + 25.50)/2 = 25.45\,cm^3$. Average titre values are normally quoted to 1 decimal place and the average should be written as $25.5\,cm^3$.

7. **(a)**

Titration	Rough	1	2	3
Final volume/cm³	24.0	23.60	23.90	23.60
Initial volume/cm³	0	0.10	0.00	0.00
Volume used/cm³	24.0	23.50	23.90	23.60

   (b) Mean titre — use values from results 1 and 3 to give $23.55\,cm^3$ (which should be written as $23.6\,cm^3$).

8. Relative molecular mass is the weighted mean mass of a molecule compared with 1/12 of the mass of a carbon-12 atom.
   Atomic number is the number of protons in the nucleus of an atom of a particular element.
   Mass number is the number of protons plus the number of neutrons in the nucleus of an atom.

9. $(NH_4)_3PO_4 = 149$
   $CH_3C_6H_2(NO_2)_3 = 227$

10. $[(54 \times 2) + (56 \times 42) + (57 \times 1)]/45 = 55.9$

11. $[(23 \times 12) + (24 \times 81) + (26 \times 7)]/100 = 24.02$ (allow variation on % of +/− 1% but the three percentage values must add up to 100%)

12. Na_2O, $Ca(OH)_2$, $(NH_4)_2SO_4$, $Al(NO_3)_3$, $ZnCO_3$

13. %H = $100 - 90.6 = 9.4$
    Ratio C : H = $(90.6/12) : (9.4/1) = 7.55 : 9.4 = (7.55/7.55) : (9.4/7.55) = 1 : 1.25 = 4 : 5$
    Empirical formula = C_4H_5 which has a mass = $48 + 5 = 53$
    Hence molecular formula = C_8H_{10} (molar mass = $106\,g\,mol^{-1}$)

14. mass of $HOC(CH_2)_2(COOH)_3.H_2O = 210$
    mass of $H_2O = 18$
    % H_2O = $(18/210) \times 100 = 8.57\%$

15. **(a)** $M = m/n = 9.0/0.05 = 180\,g\,mol^{-1}$
    (b) $M = 46 + 12 + 48 + 180 = 286\,g\,mol^{-1}$, $n = m/M = 14.30/286 = 0.05\,mol$

16. **(a)** $1220\,cm^3$
    (b) $960\,cm^3$

17. **(a)** $0.4\,mol\,dm^{-3}$
    (b) $0.04\,mol\,dm^{-3}$

18. $MgCO_3$ solid would react and a colourless solution would be formed alongside bubbles of CO_2 gas.
    (a) Moles of $MgCO_3(s) = 0.843/84.3 = 0.01\,mol$
       Therefore $0.02\,mol$ HCl(aq) is required
       volume of HCl = $0.02/0.05 = 0.4\,dm^3 = 400\,cm^3$
    (b) $0.01\,mol$ CO_2 produced, hence volume of CO_2 = $0.01 \times 24\,000 = 240\,cm^3$

19. moles of Cu = $5 \times 10^{-4}\,mol = 0.03175\,g$

20. moles of ethanol used = $5/46 = 0.11$
    moles of ethanoic acid used = $8/60 = 0.133\,mol$
    Therefore, moles of ester that could be produced = $0.11\,mol$ (theoretical yield)
    moles of ester actually produced = $7.12/88 = 0.081\,mol$
    percentage yield = $(0.081/0.11) \times 100 = 74\%$

21. atom economy = $[46/(46 + 102.90)] \times 100 = 30.9\%$

22. An acid is a proton donor. A base is a proton acceptor. A salt is formed when one or more H^+ ions in an acid are replaced by metal ions or ammonium ions.

23. **(a)** K_3PO_4 $Ba_3(PO_4)_2$
    (b) $CH_3CH_2COO^-Na^+$ $(CH_3CH_2COO^-)_2Ca^{2+}$

24 (a) Acid + metal: metal reacts/dissolves and bubbles (of H_2) are evolved

(b) Acid + carbonate: carbonate solid reacts/dissolves and bubbles (of CO_2) are evolved

(c) Acid + base: metal oxide solid reacts/dissolves (and a colourless solution is formed)

(d) Acid + alkali: no observations (colourless solutions react to form different colourless solutions)

25 (a) $H_2SO_4(aq) + Li_2O(s) \rightarrow Li_2SO_4(aq) + H_2O(l)$
$2H^+(aq) + Li_2O(s) \rightarrow 2Li^+(aq) + H_2O(l)$

(b) $3HNO_3(aq) + Al(OH)_3(aq) \rightarrow Al(NO_3)_3(aq) + 3H_2O(l)$
$H^+(aq) + OH^-(aq) \rightarrow H_2O(l)$

(c) $2HCl(aq) + BaCO_3(s) \rightarrow BaCl_2(aq) + CO_2(g) + H_2O(l)$
$2H^+(aq) + CO_3^{2-}(aq) \rightarrow CO_2(g) + H_2O(l)$

26 (a) (i) $12.70\,cm^3$
   (ii) $23.10\,cm^3$
   (iii) $9.70\,cm^3$
   (iv) $3.60\,cm^3$
   (all $+/- 0.05\,cm^3$)

(b) (i) 0.39%
   (ii) 0.21%
   (iii) 0.52%
   (iv) 1.39%

(c) (i) 10.4(0)
   (ii) 0.96%

27 mole ratio $HCl:Na_2CO_3$ is 1:2, therefore 0.0132 mol of Na_2CO_3 required.

28 $Na_2\underline{S}O_3 : S = +4$ sodium sulfate(ıv)
$\underline{Cl}O_4^- : Cl = +7$ chlorate(vıı)
$\underline{Cr}O_4^{2-}: Cr = +6$ chromate(vı)

29 $Mg + H_2SO_4 \rightarrow MgSO_4 + H_2$
Mg has been oxidised as its oxidation number changed from 0 to +2 and H has been reduced as its oxidation number went from +1 to 0.
$Mg(OH)_2 + H_2SO_4 \rightarrow MgSO_4 + 2H_2O$
This is not a redox reaction as the oxidation state of each element remains unchanged. It is in fact a neutralisation reaction.

30 $Ca^{3+}(g) \rightarrow Ca^{4+}(g) + e^-$

31 The element is silicon as there is a very large increase in ionisation energies after the 4th IE. This shows that there are four electrons in the outer shell hence the element must be in group 4.

32 $_{24}Cr = 1s^2\,2s^2\,2p^6\,3s^2\,3p^6\,3d^5\,4s^1$
$_{23}V^{2+} = 1s^2\,2s^2\,2p^6\,3s^2\,3p^6\,3d^3$

33 (a) (i) ionic
   (ii) covalent
   (iii) metallic
   (iv) covalent
   (v) ionic

(b)

34

35

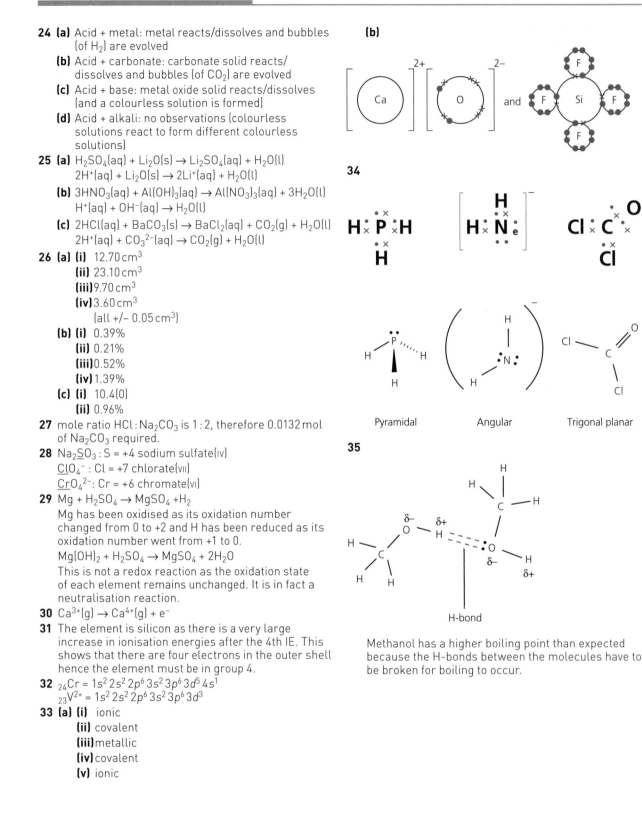

Pyramidal   Angular   Trigonal planar

H-bond

Methanol has a higher boiling point than expected because the H-bonds between the molecules have to be broken for boiling to occur.

Note: **bold** page numbers indicate defined terms.

A

accuracy and precision 10–11, 12, 18
acid–base titrations 36–39
acids 34–39
    acid and alkali reactions 35
    acid and base reactions 35
    acid and carbonate reactions 35
    acid and metal reactions 35
    acid–base titrations 36–39
    and bases 34
    neutralisation 35
    short response questions 78–82
A-level exam
    approaching the exam 53
    exam advice 55
    extended response questions 85–90
    multiple-choice questions 56–58
    question types 53–54
    short response questions 59–85
    study skills and revision techniques 6–7
    terms used in exam questions 54–55
alkalis 34, 35
amount of substance 24–33
    anhydrous salts, hydrated salts and water of crystallisation 25–26
    determination of formulae 25–26
    ideal gas equation 29–30
    mole calculations for gases 27–28
    mole calculations for solutions 28
    moles 24
    percentage yields and atom economy 32–33
    reacting masses, gas volumes and mole concentrations 26–27
    using equations 31–32
analysis 12–16
anhydrous salts 25–26
anions (negative ions) 19, 24

apparatus, choice of 10, 11, 18, 62–66
approaching the exam 53
AS exam
    approaching the exam 53
    exam advice 55
    extended response questions 85–90
    multiple-choice questions 56–58
    question types 53–54
    short response questions 59–85
    study skills and revision techniques 6–7
    terms used in exam questions 54–55
atom economy calculations 33
atomic number **19**
atomic orbitals 45
atomic radius 43
atomic structure and isotopes 19–22
Avogadro's constant 24
axes 15, 16

B

bases
    acid–base titrations 36–39
    and acids 34, 35
    short response questions 78–82
bonding and structure 47–52
    dot-and-cross diagrams 48
    electronegativity and bond polarity 49–50
    extended response questions 85–88
    intermolecular forces 50–51
    shapes of molecules and ions 48–49
    short response questions 66–68, 74–78
    structure, bonding and physical properties 51–52
    types of bond 47
bond polarity 50
burettes 11, 12, 18, 36, 37

C

carbonate and acid reactions 35
cations (positive ions) 19, 24
chemicals, safety risks 8–9
choice of apparatus 10, 11, 18, 62–66
compounds, formulae and equations 23–24
coordinate bonds **47**
covalent bonds 47, 48

D

data accuracy 18
dative covalent bonds **47**, 48
d-block elements 46
decimal places 12
density 13
dependent variable 16
determination of formulae 25–26
dipoles 49, 50, 51
dot-and-cross diagrams 48
drawing graphs 14–16
drawing tangents 15

E

electron configuration 46, 59–62
electronegativity **49**, 50, 68–70
electron-pair repulsion theory 48–49
electron shielding 43
electron structure 43–46
    atomic orbitals 45
    atomic structure and isotopes 19
    electron configuration 46
    energy levels, shells and sub-shells 43–45
    ionisation energy 43
empirical formula **25**
energy levels 43–44
enthalpy experiments 17, 18
equations 23–24, 31–32
equipment, choice of 10, 11, 18, 62–66

Index

errors
  limitations in experimental
      procedure 17
  maximum error 10–11, 12, 18,
      36, 37
  percentage error 11, 36, 37
'essay' questions 54
evaluation of experiments 17–18
exam
  approaching the exam 53
  exam advice 55
  extended response questions
      85–90
  multiple-choice questions
      56–58
  question types 53–54
  short response questions 59–85
  study skills and revision
      techniques 6–7
  terms used in exam
      questions 54–55
experiments
  improving experimental design 18
  limitations in experimental
      procedure 17
  planning 8–9, 71–74
  safety risks 8–9
extended response questions 85–90
  bonding 85–88
  question types 54
  shapes of molecules 88–89
  unstructured mole
      calculation 89–90

F
first ionisation energy **43**
formulae 23–24, 25–26

G
gas constant 29
gases
  ideal gas equation 29–30
  improving experimental
      design 18

limitations in experimental
      procedure 17
  mole calculations 27–28
  volume of 29–30
giant ionic structures 51
graphs, drawing 14–16

H
handling substances 17
hazard warning labels 9
heating substances 17
hydrated salts 25–26
hydrogen bonding 51, 68–70

I
ideal gas **29**
ideal gas equation 29–30
identifying practical techniques and
      equipment 9–10
implementing scientific
      investigations 10–12
improving experimental design 18
independent variable 16
index (standard) notation 13, 14
induced dipole–dipole
      interactions 50–51
intermolecular forces **50**, 51
interpretation of results 71–74
ionic bonds 47, 48
ionic equations 78–82
ionisation energy
  electron structure 43–44
  short response questions 59–62
ions 19, 48–49
isotopes **19**, 59–62

K
knowledge check answers 91–92

L
laboratory techniques 9–10
limitations in experimental
      procedure 17
line of best fit 15

M
mass number **19**
mass spectra 21, 22
maximum error
  implementing scientific
      investigations 10–11, 12
  precision and accuracy 18
  volumetric equipment 36, 37
measurements accuracy and
      precision 10–11, 12, 18
measuring cylinders 11
meniscus 11, 36, 37, 38
metal and acid reaction 35
metallic bonds 47
Module 1: practical skills
      assessed in the written
      examination 8–18
  analysis 12–16
  evaluation 17–18
  identifying practical techniques
      and equipment 9–10
  implementing 10–12
  planning 8–9
Module 2: foundations in
      chemistry 19–52
  acids 34–39
  amount of substance 24–33
  atomic structure and
      isotopes 19–22
  bonding and structure 47–52
  compounds, formulae and
      equations 23–24
  electron structure 43–46
  redox 39–42
molar gas volume 27
molar mass 24, 26, 27
mole calculations
  extended response questions
      89–90
  for gases 27–28
  moles from mass 26–27
  for solutions 28
  using equations 31–32

molecular formula **25**

molecules, shapes of 48–49, 88–89

moles

   definition **24**

   short response questions 62–66

multiple-choice questions 53–54, 56–58

N

negative ions (anions) 19, 24

neutralisation 35

neutrons 19

notes, organising 6

nuclear charge 43

nucleon number **19**

nucleus 19, 43, 44, 45, 47

O

OILRIG acronym 39

orbitals **45**, 46

organising your notes 6

organising your time 6–7

oxidation 39–42

oxidation number 40–41, 42

P

p-block elements 46

percentage error **11**, 36, 37

percentage yield calculations 32–33

periodic table 23, 46

permanent dipole–dipole interactions 50

physical properties of structures 51–52

pipettes 11, 36–37

planning an experiment 8–9, 71–74

polarity 50

positive ions (cations) 19, 24

practical skills 8–18

   analysis 12–16

   evaluation 17–18

identifying practical techniques and equipment 9–10

implementing 10–12

planning 8–9

short response questions 82–85

precision 10–11, 12, 18

proton number **19**

protons 19, 34, 43

Q

qualitative data 12

quantitative data 12

questions and answers 53–92

   approaching the exam 53

   exam advice 55

   multiple-choice questions 56–58

   overview 55

   question types 53–54

   short response questions 59–85

   terms used in exam questions 54–55

R

reacting masses 26–27, 31

recording results 12

redox 39–42

   oxidation 39–42

   oxidation number 40–41

   redox reactions 41–42

   short response questions 66–68

reduction 41, 42

relative atomic mass 19–22, 25

relative formula mass 20, 21

relative isotopic mass 19–20

relative mass 19–22

relative molecular mass 20, 21, 24, 25

results

   interpretation of 71–74

   recording 12

revision techniques 6–7

risk assessment 8–9

S

safety risks 8–9

salts 25–26, 35

s-block elements 46

scientific investigations *see* experiments

second ionisation energy **43**

shapes of molecules and ions 48–49, 88–89

shells 43–44

shielding 43

short response questions 59–85

   acids, bases and ionic equations 78–82

   bonding and related properties 74–78

   electronegativity and hydrogen bonding 68–70

   isotopes, electron configuration and ionisation energy 59–62

   moles and choice of apparatus 62–66

   planning an experiment and interpretation of results 71–74

   practical skills in chemistry 82–85

   question types 54

   redox, bonding and structure 66–68

significant figures 12–13

simple covalent structures 52

simple molecular lattices 52

solutions 17, 28

standard (index) notation 13, 14

standard solutions **38**

strong acids 34

structure and bonding 51–52, 66–68

study skills 6–7

sub-shells 44–45

substances, handling and heating 17

Index

T

tangents, drawing 15
techniques, practical 9–10
terms used in exam questions 54–55
theoretical yield **32**
time, organising
    for exam 6–7
    experiments 17
titrations 36–39

U

unstructured mole calculations
    89–90
using equations 31–32

V

'valency cross-over' technique 23
volume of gases 29–30
volumetric analysis 36–39

volumetric equipment 11,
    36–38
volumetric flasks 11, 36, 37–38

W

warning labels 9
water of crystallisation 25
weak acids 34
written examination *see* exam